RENEWALS 458-4574
DATE DUE

WITHDRAWN
UTSA Libraries

Valuation of Ecological Resources

Integration of Ecology and Socioeconomics
in Environmental Decision Making

Other Titles from the Society of Environmental Toxicology and Chemistry (SETAC):

Ecosystem Responses to Mercury Contamination: Indicators of Change
Harris, Krabbenhoft, Mason, Murray, Reash, Saltman, editors
2007

Genomic Approaches for Cross-Species Extrapolation in Toxicology
Benson and Di Giulio, editors
2007

*New Improvements in the Aquatic Ecological Risk Assessment
of Fungicidal Pesticides and Biocides*
Van den Brink, Maltby, Wendt-Rasch, Heimbach, Peeters, editors
2007

Freshwater Bivalve Ecotoxicology
Farris and Van Hassel, editors
2006

*Estrogens and Xenoestrogens in the Aquatic Environment:
An Integrated Approach for Field Monitoring and Effect Assessment*
Vethaak, Schrap, de Voogt, editors
2006

*Assessing the Hazard of Metals and Inorganic Metal Substances
in Aquatic and Terrestrial Systems*
Adams and Chapman, editors
2006

Perchlorate Ecotoxicology
Kendall and Smith, editors
2006

Natural Attenuation of Trace Element Availability in Soils
Hamon, McLaughlin, Stevens, editors
2006

*Mercury Cycling in a Wetland-Dominated Ecosystem:
A Multidisciplinary Study*
O'Driscoll, Rencz, Lean
2005

For information about SETAC publications, including SETAC's international journals, *Environmental Toxicology and Chemistry and Integrated Environmental Assessment and Management*, contact the SETAC Administrative Office nearest you:

SETAC Office
1010 North 12th Avenue
Pensacola, FL 32501-3367 USA
T 850 469 1500   F 850 469 9778
E setac@setac.org

SETAC Office
Avenue de la Toison d'Or 67
B-1060 Brussels, Belgium
T 32 2 772 72 81   F 32 2 770 53 86
E setac@setaceu.org

www.setac.org
Environmental Quality Through Science®

Valuation of Ecological Resources

Integration of Ecology and Socioeconomics in Environmental Decision Making

Ralph G. Stahl, Jr. | Lawrence A. Kapustka
Wayne R. Munns, Jr. | Randall J. F. Bruins

From the Society of Environmental Toxicology and Chemistry workshop on
Valuation of Ecological Resources:
Integration of Ecological Risk Assessment and Socio-Economics to
Support Environmental Decisions

Pensacola, Florida, USA
4-9 October 2003

Coordinating Editor of SETAC Books
Joseph W. Gorsuch
Gorsuch Environmental Management Services, Inc.
Webster, New York, USA

CRC is an imprint of the Taylor & Francis Group,
an informa business

Published in collaboration with the Society of Environmental Toxicology and Chemistry (SETAC)
1010 North 12th Avenue, Pensacola, Florida 32501
Telephone: (850) 469-1500 ; Fax: (850) 469-9778; Email: setac@setac.org
Web site: www.setac.org
ISBN: 978-1-880611-15-9 (SETAC Press)

© 2008 by the Society of Environmental Toxicology and Chemistry (SETAC)
SETAC Press is an imprint of the Society of Environmental Toxicology and Chemistry.

No claim to original U.S. Government works
Printed in the United States of America on acid-free paper
10 9 8 7 6 5 4 3 2 1

International Standard Book Number-13: 978-1-4200-6262-5 (Hardcover)

This book contains information obtained from authentic and highly regarded sources. Reprinted material is quoted with permission, and sources are indicated. A wide variety of references are listed. Reasonable efforts have been made to publish reliable data and information, but the author and the publisher cannot assume responsibility for the validity of all materials or for the consequences of their use.

The content of this publication does not necessarily reflect the position or policy of the U.S. government or sponsoring organizations and an official endorsement should not be inferred.

No part of this book may be reprinted, reproduced, transmitted, or utilized in any form by any electronic, mechanical, or other means, now known or hereafter invented, including photocopying, microfilming, and recording, or in any information storage or retrieval system, without written permission from the publishers.

For permission to photocopy or use material electronically from this work, please access www.copyright.com (http://www.copyright.com/) or contact the Copyright Clearance Center, Inc. (CCC) 222 Rosewood Drive, Danvers, MA 01923, 978-750-8400. CCC is a not-for-profit organization that provides licenses and registration for a variety of users. For organizations that have been granted a photocopy license by the CCC, a separate system of payment has been arranged.

Trademark Notice: Product or corporate names may be trademarks or registered trademarks, and are used only for identification and explanation without intent to infringe.

Library of Congress Cataloging-in-Publication Data
Valuation of ecological resources : integration of ecology and socioeconomics in environmental decision making / Ralph G. Stahl Jr. ... [et al.].
p. cm.
Includes bibliographical references and index.
ISBN-13: 978-1-4200-6262-5 (alk. paper)
1. Environmental risk assessment. 2. Ecological risk assessment. 3. Environmental management. I. Stahl, Ralph G., 1953-
GE145.V365 2008
333.71'4--dc22   2007019369

Visit the Taylor & Francis Web site at
http://www.taylorandfrancis.com

and the CRC Press Web site at
http://www.crcpress.com

and the SETAC Web site at
www.setac.org

SETAC Publications

Books published by the Society of Environmental Toxicology and Chemistry (SETAC) provide in-depth reviews and critical appraisals on scientific subjects relevant to understanding the impacts of chemicals and technology on the environment. The books explore topics reviewed and recommended by the Publications Advisory Council and approved by the SETAC North America, Latin America, or Asia/Pacific Board of Directors; the SETAC Europe Council; or the SETAC World Council for their importance, timeliness, and contribution to multidisciplinary approaches to solving environmental problems. The diversity and breadth of subjects covered in the series reflect the wide range of disciplines encompassed by environmental toxicology, environmental chemistry, and hazard and risk assessment, and life-cycle assessment. SETAC books attempt to present the reader with authoritative coverage of the literature, as well as paradigms, methodologies, and controversies; research needs; and new developments specific to the featured topics. The books are generally peer reviewed for SETAC by acknowledged experts.

SETAC publications, which include Technical Issue Papers (TIPs), workshops summaries, newsletter (*SETAC Globe*), and journals (*Environmental Toxicology and Chemistry* and *Integrated Environmental Assessment and Management*), are useful to environmental scientists in research, research management, chemical manufacturing and regulation, risk assessment, and education, as well as to students considering or preparing for careers in these areas. The publications provide information for keeping abreast of recent developments in familiar subject areas and for rapid introduction to principles and approaches in new subject areas.

SETAC recognizes and thanks the past coordinating editors of SETAC books:

C.G. Ingersoll, Columbia Environmental Research Center
US Geological Survey, Columbia, Missouri, USA

T.W. La Point, Institute of Applied Sciences
University of North Texas, Denton, Texas, USA

B.T. Walton, US Environmental Protection Agency
Research Triangle Park, North Carolina, USA

C.H. Ward, Department of Environmental Sciences and Engineering
Rice University, Houston, Texas, USA

Contents

List of Figures ix
List of Tables xi
About the Editors xiii
Contributors xvii
Preface xix
Acknowledgments xxi

Chapter 1 Approaching the Problem of Ecological Valuation............................ 1

  Ralph G. Stahl, Jr., Larry Kapustka, Randall J.F. Bruins, and Wayne R. Munns, Jr.

Chapter 2 Sociocultural Valuation of Ecological Resources 9

  Jessica Glicken Turnley, Michael D. Kaplowitz, Orie L. Loucks, Beth L. McGee, and Thomas Dietz

Chapter 3 Integrating Economics and Ecological Assessment.......................... 45

  Richard C. Bishop, Joshua Lipton, Michael Margolis, Norman Meade, George L. Peterson, and Alan Randall

Chapter 4 Valuation Methods 59

  W.L. (Vic) Adamowicz, David Chapman, Gene Mancini, Wayne R. Munns, Jr., Andrew Stirling, and Ted Tomasi

Chapter 5 Complexity in Ecological Systems 97

  Katherine von Stackelberg, Sam Luoma, Ron McCormick, Kristin Skrabis, Elaine Dorward-King, and Stephen Polasky

Chapter 6 Organizing and Integrating the Valuation Process 129

  Brian Heninger, Greg Biddinger, Chester Joy, Catherine L. Kling, Doug Reagan, and Travis S. Schmidt

Chapter 7 Synthesis, Recommendations, and Conclusions 165

Randall J.F. Bruins, Wayne R. Munns, Jr., Larry Kapustka, and Ralph G. Stahl, Jr.

Case Study 1 National Park Establishment, Philippines 173

Doug Reagan

Case Study 2 Large PCB-Contaminated River under Superfund Interface between Remediation and Restoration 179

Katrina Von Stackleberg

Case Study 3 Extractive Development and the Valuation of Biodiversity and Community Needs in Madagascar 187

Elaine Dorward-King

Case Study 4 1991 Gulf War Oil Spill Long-Term Impacts on Shoreline Habitats 195

Jacqui Michel

Case Study 5 Example of Valuing the Ecological Benefits from the Clean Air Act and 1990 Amendments 201

Brian T. Heninger

Case Study 6 The CALFED Bay–Delta Program — A Case Study in Intertwined Ecological and Environmental Resource Issues 207

Samuel N. Luoma

Index 217

List of Figures

Figure 1.1	Inputs to the ecological risk management decision.	2
Figure 1.2	Relationship between risk tolerance and the perceived value (in terms of function, scarcity, uniqueness, inaccessibility, etc.) of an ecological resource.	3
Figure 2.1	The context of behavior after Spulak and Turnley (2005).	13
Figure 2.2	Selecting social indicators.	29
Figure 4.1	Contrasting degrees of incomplete knowledge (Stirling 1999)	79
Figure 4.2	Analysis approaches used by economists and ecologists.	86
Figure 6.1	Food web for a forest ecosystem (modified from Reagan 2006).	153
Figure C1.1	Environmental service flows for indigenous communities.	175
Figure C6.1	Population size (number of adult individuals) of winter-run Chinook salmon in the Sacramento River watershed	209

List of Tables

Table 1.1	Typical questions and issues that may arise from respective groups involved with ecological risk management decisions	4
Table 2.1	Social indicators in data collection for a social impact assessment	28
Table 4.1	An assessment of valuation methods	75
Table 4.2	Decision-making procedure	92
Table 5.1	Integration of ecology and economics	108
Table 6.1	Spatial and temporal scale of environmental decisions in a valuation context	140
Table 6.2	Examples of single-factor and multifactor decision focuses	141
Table 6.3	Functional components of the tropical rain forest ecosystem in New Guinea	153
Table 6.4	Ecologically relevant attributes of the functional components identified for a tropical rain forest ecosystem in New Guinea	155
Table C6.1	Trade-offs in the CALFED Bay–Delta Program: Examples of actions and perceived net cost or benefit of each	211

About the Editors

Ralph G. Stahl, Jr., PhD, DABT, a native of Houston, Texas, received his BS in marine biology from Texas A&M University (cum laude) in 1976, his MS in biology from Texas A&M University in 1980, and his PhD in environmental science and toxicology from the University of Texas School of Public Health in 1982. He then became a senior postdoctoral fellow in the Department of Pathology at the University of Washington in Seattle, where he investigated the impact of genetic toxins on biological systems. Stahl joined the DuPont Company in 1984 and in the intervening years has held both technical and management positions in the research and internal consulting arenas. His research over the last 24 years has focused primarily on evaluating the effects of chemical stressors on aquatic and terrestrial ecosystems. Since 1993, Stahl has been responsible for leading DuPont's corporate efforts in ecological risk assessment and natural resource damage assessments for site remediation.

He has been involved with oceanographic studies in the Atlantic Ocean, Pacific Ocean, Gulf of Mexico, and Caribbean Sea; biological and ecological assessments at contaminated sites in the United States, Europe, and South America; and numerous toxicological studies with mammals, birds, and aquatic organisms. He has been named by the US Environmental Protection Agency (USEPA), Army Corps of Engineers, Strategic Environmental Research and Development Program, National Institutes of Environmental Health Sciences, National Academy of Science, Water Environment Research Foundation, National Oceanographic and Atmospheric Administration (NOAA), State of Washington, State of Texas, and others to national or state peer review panels on ecological risk assessment, endocrine disruption in wildlife, or natural resource injury determination.

Stahl is a member of the USEPA's Science Advisory Board (Advisory Council on Clean Air Compliance Analysis, Ecological Effects Subcommittee) and is active in the Society of Environmental Toxicology and Chemistry (SETAC), serving on the Ecological Risk Assessment Advisory Group. He is board certified in general toxicology and is a diplomate of the American Board of Toxicology. He has authored more than 30 peer-reviewed publications on topics in environmental toxicology, ecological risk assessment, and risk management. He recently edited 2 books and is currently coediting a third book stemming from a SETAC Education Foundation-sponsored workshop on the valuation of ecological resources.

Stahl currently resides in Wilmington, Delaware, where in his spare time he enjoys tying flies, fly-fishing, and watching his teenage son play soccer.

Lawrence A. (Larry) Kapustka is a senior ecotoxicologist and associate with Golder Associates Ltd. in Calgary, Alberta. His work focuses on using spatially explicit risk assessments, integrating environmental assessment practices with environmental management decision processes, and advancing the emerging methods in the field of ecological valuation. In the 15 years prior to joining Golder in 2005, Kapustka was president of Ecological Planning and Toxicology, Inc., in Corvallis, Oregon. From 1988 through 2000, he was leader of the Hazardous Waste and Plant Toxicology Teams at the USEPA Laboratory in Corvallis, Oregon. From 1978 through 1988, Kapustka was a professor of botany at Miami University, Oxford, Ohio, where he taught various ecology courses and conducted ecological research exploring dinitrogen fixation processes, mycorrhizal functions, fire ecology, and allelopathic interactions. He worked for 3 years before that at the University of Wisconsin–Superior on the role of vegetation in suppressing erosion. Kapustka earned his PhD in botany from the University of Oklahoma in Norman in 1975. He received his MS (1972) and BS Ed (1970) from the University of Nebraska–Lincoln. Over his career, he has published more than 65 peer-reviewed journal articles, 16 book chapters, 3 editions of an environmental science textbook for nonmajors, and more than 100 technical reports.

Kapustka is active in several professional societies. He is a certified senior ecologist through the Ecological Society of America (ESA) and is a member of the International Association of Landscape Ecologists (IALE). In the Society of Environmental Toxicology and Chemistry (SETAC), Larry serves as the chair of the Ecological Risk Assessment Advisory Group (ERAAG), which has sponsored Pellston Workshops on populations and ecological valuation. He serves on the American Society for Testing and Materials (ASTM) Executive Subcommittee ASTM E-47 (Biological Effects and Environmental Fate), and chairs the subcommittee E47.02 (Terrestrial Toxicology and Assessment). He was appointed to serve on ASTM's Committee on Technical Committee Operations (COTCO), which is composed of 2 Technical Operations (01) and Regulations (02) subcommittees.

Wayne R. Munns, Jr., PhD, is the associate director for science for the USEPA Atlantic Ecology Division, Office of Research and Development, in Narragansett, Rhode Island. A marine ecologist by training (University of Rhode Island, 1984), he has expertise in developing and applying quantitative methods for ecological risk assessment, ecological modeling with particular emphasis on population dynamics, and large spatial scale environmental assessments. He has conducted research and managed research programs addressing ocean disposal, hazardous waste sites, contaminated sediments, wildlife risk assessment, and environmental criteria development. His current interests include the development of population-level ecological risk assessment methods to support regulatory decisions, ecological services and their valuation, and the integration of assessment approaches to enhance the value of information supporting environmental protection decisions. Prior to joining the USEPA, Munns was a senior scientist, division manager, and assistant vice president for Science Applications International Corporation. He has been a member of the USEPA's Risk Assessment Forum and has advised the World Health Organization on the integration of human health and ecological risk assessment. He has also contributed to the development of several previous Pellston Workshops as a steering committee member, editor, and participant. He is a past member of the editorial board for SETAC's *Environmental Toxicology and Chemistry* series, and currently serves as vice chair of SETAC's Ecological Risk Assessment Advisory Group.

Randall J.F. Bruins is an environmental scientist in the USEPA's Office of Research and Development, Cincinnati, Ohio. He received his BS (1978) and MS (1980) in zoology from Miami University and his PhD (1997) in environmental science from Ohio State University. His dissertation research examined methods for reducing flooding in central China through ecological strategies such as the replacement of low-lying rice with native wetland crops. His USEPA research has focused on methods for integrating ecological risk assessment and economic analysis, and in 2005 he coedited (with Matthew Heberling) a book on this subject, *Economics and Ecological Risk Assessment: Applications to Watershed Management*. Bruins also is a coauthor of the USEPA's draft strategy for measuring the benefits of ecosystem protection (the *Ecological Benefits Assessment Strategic Plan*). His current position is in the USEPA's National Exposure Research Laboratory, where he heads the Ecosystems Research Branch, a group that develops methods and indicators for monitoring the ecological condition of streams, rivers, and wetlands.

Contributors

W.L. (Vic) Adamowicz
University of Alberta
Edmonton, Alberta, Canada

Gregory R. Biddinger
ExxonMobil Biomedical Sciences, Inc.
Houston, Texas, USA

Richard C. Bishop
University of Wisconsin
Madison, Wisconsin, USA

Randall J.F. Bruins
US Environmental Protection Agency
Cincinnati, Ohio, USA

David Chapman
Stratus Consulting
Boulder, Colorado, USA

Thomas Dietz
Michigan State University
East Lansing, Michigan, USA

Elaine J. Dorward-King
Rio Tinto plc
London, United Kingdom

Brian T. Heninger
US Environmental Protection Agency
Washington DC, USA

Chester Joy
US Government Accountability Office
Washington DC, USA

Michael D. Kaplowitz
Michigan State University
East Lansing, Michigan, USA

Larry Kapustka
Golder Associates Ltd.
Calgary, Alberta, Canada

Catherine L. Kling
Iowa State University
Ames, Iowa, USA

Joshua Lipton
Stratus Consulting, Inc.
Boulder, Colorado, USA

Orie L. Loucks
Miami University
Oxford, Ohio, USA

Samuel N. Luoma
US Geological Survey
Menlo Park, California, USA

Gene Mancini
E.R. Mancini & Associates
Camarillo, California, USA

Michael Margolis
Universidad de Guanajuato
Guanajuato, México

Ronald J. McCormick
Golder Associates Ltd.
Calgary, Alberta, Canada

Beth L. McGee
Chesapeake Bay Foundation
Annapolis, Maryland, USA

Norman F. Meade
National Oceanic and Atmospheric
  Administration
Silver Spring, Maryland, USA

Jacqueline Michel
Research Planning, Inc.
Columbia, South Carolina, USA

Wayne Munns
US Environmental Protection Agency
Narragansett, Rhode Island, USA

George Peterson, retired
Forest Service, USDA
Fort Collins, Colorado, USA

Steve Polasky
University of Minnesota
St. Paul, Minnesota, USA

Alan Randall
Ohio State University
Columbus, Ohio, USA

Douglas P. Reagan
Doug Reagan & Associates
Castle Rock, Colorado, USA

Travis S. Schmidt
Colorado State University
Fort Collins, Colorado, USA

Kristin Skrabis
US Department of the Interior
Washington DC, USA

Ralph G. Stahl, Jr.
DuPont Corporate Remediation Group
Wilmington, Delaware, USA

Andrew Stirling
University of Sussex
Brighton, East Sussex, United Kingdom

Ted Tomasi
Entrix Inc.
New Castle, Delaware, USA

Jessica Glicken Turnley
Galisteó Consulting Group, Inc.
Albuquerque, New Mexico, USA

Katherine von Stackelberg
Harvard School of Public Health
Boston, Massachusetts, USA

Preface

Risk managers and other environmental decision makers in the public and private sectors rely on diverse, and not always congruent, sets of information from which they evaluate and choose among management options. Implementing one management option may be very cost-effective, but fail to accommodate the needs or desires of some set of stakeholders. Implementing a different option may satisfy the majority of stakeholders, but at the same time may place an unacceptably high burden on an individual, a group, or perhaps the public at large. Each option, or decision, will entail trade-offs whose consequences, in an ideal state, will be understood and balanced among the interested stakeholders, as well as among the inputs (legal, scientific, technological, etc.) informing them.

A workshop organized by the Society of Environmental Toxicology and Chemistry (SETAC) in October 2003 explored the state of the science and opportunities for incorporating ecology and socioeconomics in environmental decision-making processes. The workshop considered 1) the roles of ecology, sociology, and economics in the environmental decision-making process; 2) the scientific underpinnings for, and the roles of society and regulatory and legislative bodies in, determining the "value" of complex ecological resources; 3) alternatives to the monetary valuation of ecological resources that can enhance environmental decision-making processes; and 4) ecological and economic information sets, tools, and analytical frameworks needed to make sound environmental decisions. This book summarizes the major conclusions reached during the workshop and sets the stage for further dialogue within the environmental decision-making community.

Ralph G. Stahl, Jr.

Lawrence A. Kapustka

Wayne R. Munns, Jr.

Randall J.F. Bruins

Acknowledgments

The authors and editors of this book wish to acknowledge the following sponsors of a Society of Environmental Toxicology and Chemistry (SETAC) Pellston Workshop on Valuation of Ecological Resources: Integration of Ecological Risk Assessment and Socio-Economics to Support Environmental Decisions, 4–9 October 2003, Pensacola, Florida:

American Chemistry Council
American Petroleum Institute
E.I. du Pont de Nemours and Company
Electric Power Research Institute
ENVIRON International Corporation
ExxonMobil
National Oceanic and Atmospheric Administration
Rio Tinto plc
URS Corporation
US Environmental Protection Agency
US Geological Survey/CALFED

This book is the product of the workshop and subsequent efforts.

We also wish to thank Professor Darrell Bosch, Department of Agricultural and Applied Economics, Virginia Tech, for his expertise in reviewing and suggesting refinements to the manuscript.

The opinions expressed in this book are those of the participants and may not reflect those of any of their agencies, the funding agencies, or SETAC.

1 Approaching the Problem of Ecological Valuation

*Ralph G. Stahl, Jr, Larry Kapustka,
Randall J.F. Bruins, and Wayne R. Munns, Jr.*

CONTENTS

1.1 Introduction 1
1.2 The Challenge: Integrating Socioeconomics and Ecological
    Risk Assessment 4
1.3 A Paradigm Shift 5
References 7

1.1 INTRODUCTION

Environmental decision making is a messy business. Risk managers and other environmental decision makers in the public and private sectors often must balance the interests of a variety of stakeholders. A particular management option may be cost-effective but fail to accommodate the needs or desires of some set of stakeholders. A different option may appeal to a majority of stakeholders but place an unacceptably high burden (costs or other) on an individual, a group, or perhaps the public at large. Each option entails trade-offs with consequences that in an ideal state are understood and balanced. In this ideal, options would be selected that best reflect stakeholder desires and conform to scientific, technological, social, and legal facts and constraints (Figure 1.1). It is clear that to maximize effectiveness, the decision-making process should be supported by concepts, tools, and information that readily bring together all of the factors that enter into the decision.

The modern environmental movement that began in the late 1960s has evolved to address a wide range of human-induced influences. Efforts to improve the quality of air and water were initially focused on correcting problems that resulted from prior emissions and releases. Processes including environmental impact assessment (EIA) and risk assessment (RA) were then devised to evaluate, in advance, the possible outcomes of management actions, such as undertaking a substantive modification of the environment (such as siting a road, opening a mine, or logging a forest) or granting authorization for the manufacture and use of chemicals. The information developed by these science-based processes supports decisions to permit, deny, or otherwise regulate the anticipated action. Despite the usefulness of tools like EIA

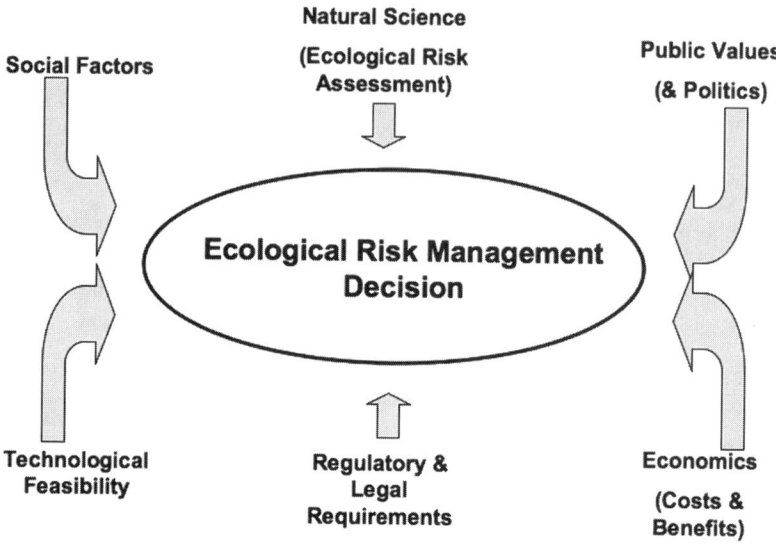

FIGURE 1.1 Inputs to the ecological risk management decision. *Source:* Redrawn from Stahl et al. (2001).

and RA, they provide but a fraction of the information necessary to support decisions concerning complex environmental issues.

Even as these science-based tools were developing, society kept its eye on the financial costs incurred under the various decision options. For example, the Inflationary Impact Statements of the mid-1970s, while emphasizing the utilitarian goal of maximizing societal welfare in environmental decision making, required that the effects of those decisions be monetized (USEPA 2000). Similarly, the federal Food and Drug Administration evaluated the trade-offs between human health and economic impacts in developing acceptable limits for polychlorinated biphenyls (PCBs) in fish (42 FR 17487). This has led to the continuing challenge of assigning value to ecological entities (species, communities, etc.) and processes. The prevalent practice is to monetize that value based on society's willingness to pay whenever practical. Yet, ascertaining the value of an environmental experience (e.g., observing an osprey capture a trout) versus that of a market-based commodity (e.g., the price per pound for hatchery trout) is difficult at best. Despite the seemingly impossible task of equating such disparate entities, pressures are growing to do so, and to do so with consistency. The national debate in the United States on the use of benefit–cost analysis in environmental decision making has waxed and waned over the past several years with little or no consensus among the various players from government, industry, the public, and other sectors.

The discord that tends to arise among economists, ecologists, risk assessors, lawyers, and the public places significant stress on the decision maker(s) and can be one of the main reasons that decisions are delayed. Seemingly endless debates can arise about whether or not it makes ecological, financial, or even legal sense to protect certain species or populations of a particular species (Eisner et al. 1995), or

Approaching the Problem of Ecological Valuation

the habitats upon which these species may depend (Nash 1991). Even identifying levels of protection to be afforded to various resources bedevils scientists and decision makers (USEPA 1997; Barton 2001). At the same time, the decision maker and those informing the decision have difficulty gauging the public's views and values, particularly when those values are based on cultural or individual beliefs rather than material gain.

Deciphering stakeholders' and the public's views on ecological resources, particularly as they pertain to decisions that might impact their ability to utilize, harvest, or otherwise benefit from those resources, is difficult (McDaniels 1995). This creates an intriguing challenge to ecologists, economists, and decision makers, as the success of a decision often hinges on an accurate portrayal of the value of ecological resources, the trade-offs to be made, and other factors that may be important to the array of stakeholders. Complicating the issue is an individual's or society's willingness to accept risk (risk tolerance), which is dynamic and changes across the range of the perceived value of the resource (Figure 1.2) as well as among cultures, geographical regions, and generations (Bingham et al. 1995).

Incorporation of values into environmental decision making is receiving increasing attention. Recently, the National Academy of Science, through its research arm, the National Research Council (NRC), was asked to undertake examination of the values associated with biodiversity (NRC 1999) and with aquatic ecosystems (NRC 2004). In addition, 2 important workshops sponsored by USEPA (USEPA 2001a, 2001b) sought to understand the public's views of environmental decision making and their involvement therein (USEPA 2001a), and to appraise how various disciplines (economics, social psychology, anthropology, and decision science) can contribute to understanding and incorporating those views (USEPA 2001b). Each of these efforts has acknowledged the scientific and social challenges underlying valuation of ecological resources, as well as the controversies that often accompany attempts

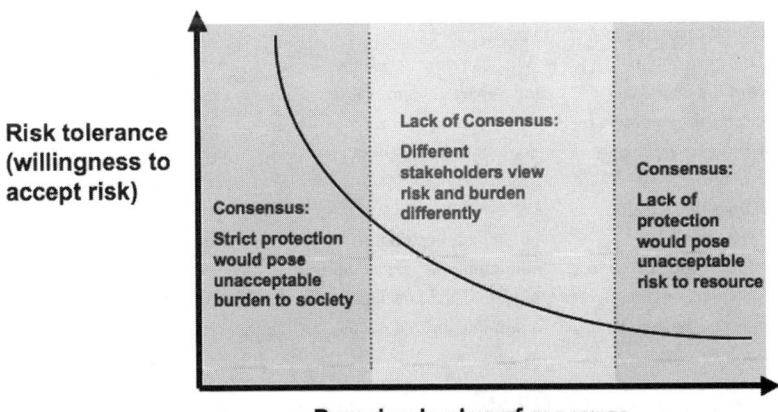

FIGURE 1.2 Relationship between risk tolerance and the perceived value (in terms of function, scarcity, uniqueness, inaccessibility, etc.) of an ecological resource. *Source:* Redrawn from Belzer (2001).

to quantify ecological value to support decision making. Despite these efforts, the issue of how best to incorporate economic, cultural, and social values into environmental decision making is as yet unresolved. This is particularly true with respect to their interactions with ecological risk assessment and risk management (Stahl et al. 2001), a key stimulus for the Society of Environmental Toxicology and Chemistry (SETAC)-sponsored workshop upon which this book is based (SETAC 2003).

1.2 THE CHALLENGE: INTEGRATING SOCIOECONOMICS AND ECOLOGICAL RISK ASSESSMENT

Improving environmental management requires the effective integration of environmental economics and ecology within a decision-making framework. Economists, ecologists, and environmental decision makers have perspectives unique to their training (Table 1.1) and historically have approached these questions in an insular manner. As observed by the NRC (2004), for example, ecologists tend to focus on

TABLE 1.1
Typical questions and issues that may arise from respective groups involved with ecological risk management decisions

Decision makers	Economists	Ecologists
• What is the context in which I am making decisions?	• What are the relationships between ecosystem function, human actions (positive or negative), and services that are valued by society?	• What ecological effects are sensitive to the stressor and relevant to management goals?
• How does that affect the data that I need to make that decision?	• What types of data are provided by biologists and ecologists? How do they compare with those I need?	• How can I establish baseline conditions (i.e., but for the actions of humans, what would the ecosystem be like)?
• At what point in the life cycle of a business or action am I operating?	• What can science tell me about the impact at various levels of biological organization, and how does that impact affect my economic analysis?	• How can information regarding ecological conditions and trends be forecast across space and time?
• Does the environmental aspect of the decision have strategic importance to my business or agency, or the public?	• How should recovery of an individual, population, or community be measured or modeled, and how can I assimilate those data into my evaluations?	• What metrics can I use to identify shifts and changes in ecological resources and processes?
• When does the decision need to be made?		• How can the relative importance of different triggers be characterized with respect to various scenarios or consequences?
• How much time do I have to collect data, and how rigorous do I have to be in getting those data?		
• How do I measure the cost-effectiveness of the action I take?	• How can I improve the institutional structure for incorporating economic science into risk management at, for example, contaminated sites?	• How can I characterize triggers with long response times (i.e., a process set in motion that may take years to be readily observable)?

ecosystem structure and function, while economists assign value to changes in ecosystem services. Yet, each also needs information provided by the other to be effective in the decision-making process. Thus, decision making could be improved if ecologists learn to characterize the capacity of ecological systems to provide the goods and services expected by humans with conflicting expectations, developing a common set of measures that facilitate valuation of those goods and services, and economists learn how to incorporate the aspects of ecological structure and function valued by ecologists into their evaluations. Environmental decisions would most benefit by explicit acknowledgment by all parties of the ecological values that stakeholders and society hold. Greater interaction and integration of the various parties to environmental decision making are needed to achieve this end. Even so, it is also evident that some decisions made by society, governments, businesses, and individuals will not have their base in monetary terms.

1.3 A PARADIGM SHIFT

Accompanying the environmental movement of the past 4 decades has been a shift in perceptions about the environment, human societies, and how humans interface with the environment. And there is greater acknowledgment of the interrelatedness among actions and consequences. These shifts in perspective can influence how environmental issues are addressed and how evaluation of the trade-offs among management options is approached. A previous SETAC/SOT (Society of Toxicology) Pellston Workshop (Di Guilio and Benson 2002) examined the broad interconnections between human and environmental well-being. A product of that workshop was a conceptual model (see Miranda et al. 2002) that described the world as made up of 2 primary systems: the "Natural System" (i.e., with elements of human and nonhuman populations, natural resources, climate, environmental stressors, etc.) and the "Social System" (i.e., with elements of economic, political, and religious systems; traditions and customs; belief systems; science and technology; and other human social constructs). It was recognized during that workshop that environmental decision making should take into account the elements and attributes of both systems to be optimally effective; yet the tools and information sets to do so generally are lacking. Several recommendations were made during that workshop to improve society's ability to make effective environmental decisions, including incorporating economic considerations in those decisions.

The view offered by the Miranda et al. (2002) conceptual model, and championed by Campbell (2001) and others, is one of integration and holism as they apply to environmental decision making. When issues are addressed in isolation (i.e., without considering the interconnectedness of complex systems), decisions are based on fragmented input. If humans are viewed as being outside of the biosphere or ecological system, actions can be taken as if there were no environmental consequences. And if so, the likelihood increases of making ineffective decisions, often with unintended consequences. Conversely, if humans are viewed as part of the ecological system, the stage is set for greater awareness of consequences. Thus, even a decision for "no action" (e.g., letting naturally occurring fires burn their natural course) will be recognized as having consequences.

As suggested earlier, this new paradigm also should integrate the perspectives of ecologists, economists, and social scientists into more holistic assessments of decision options. Collectively, these groups would work closely with risk managers during the decision-making process, being both responsive to the kinds and forms of information needed to support individual decisions and sensitive to the assumptions that they themselves bring to the assessments. This latter would be facilitated by broad discussion of the various valuation frames held by each group, including both pragmatic and theoretical elements. Such dialog can be a platform for exploring the utility of ecological valuation methods alternate to standard neoclassical monetization based on individuals' preferences, including biophysical-based methods, multi-attribute indices, and others. In doing so, various valuation schemes can be judged as to their ability to capture the complex and dynamic nature of ecological systems, as well as the interconnectedness among resources that plays out in a broader ecological context. Further, this dialog can help identify more effective ways for incorporating the cultural and sociological contexts of decisions into the risk management process.

The goal for integration of socioeconomics and ecological risk assessment is a common approach that accommodates different perspectives on ecological valuation in a cohesive, flexible framework to support sound and transparent environmental decision making. Achieving this end will require creative thinking and critical evaluation of the approaches and tools currently used. Our hope in this publication is that the views developed by the various authors reflect and foster this paradigm shift.

To achieve this, we developed a series of chapters to illustrate the various components of environmental decision making. We chose to sequence the chapters so that we introduce the concept of valuation from both a social and economic perspective, followed by methods and then a chapter on how the information is used in decision making. We open with a chapter on the sociocultural valuation of ecological resources (Chapter 2) because it explores the nature of "value" in the broadest of senses. Chapter 2 presents different concepts of value and provides examples of their use in environmental decision making, with an emphasis on the use of sociocultural assessment methods to support private sector decisions. Chapter 3 then focuses on the economist's concept of value. Because people's views of the economic approach are so disparate — with some seeing it as the only trustworthy guide to environmental decisions and others rejecting its use out of hand — Chapter 3 devotes considerable attention to the moral and philosophical bases of this approach. Chapter 4 describes and assesses the usefulness of a broad array of environmental resource valuation methods, including both economic and noneconomic methods. It provides criteria by which decision makers can evaluate the suitability of each method to a given decision context. In Chapter 5 the view momentarily shifts from valuation of ecosystems to an examination of their complex and hierarchical nature. The chapter emphasizes the need for explicit treatment of scale and uncertainty when attempting the integration of ecology and economics. Chapter 6 focuses on the needs of the decision maker. It describes the ecological and economic information sets, tools, and analytical frameworks needed to support sound environmental decisions, and discusses how these may vary according to the type of decision being made. The final chapter, Chapter 7, provides a synthesis of workshop findings and presents recommendations for further study.

We also have provided a set "case studies," international in scope, to illustrate how complex environmental decisions have been made in recent times. These include: planning the redevelopment of the former US Navy base at Subic Bay, Philippines; evaluating issues at the interface of remediation and restoration of a large PCB-contaminated river; designing ecologically sensitive approaches for mineral mining in Madagascar; determining a remediation approach for shoreline habitat impacted by oil spills during the 1991 Persian Gulf War; valuing ecological benefits from the Clean Air Act; and determining how to apportion water from California, USA's Sacramento River Delta between societal and ecological needs. It is our hope that this book will be the first of many that begin to formulate the scientific precepts necessary for improving environmental decisions worldwide.

REFERENCES

Barton AL. 2001. Deciding what resources to protect and setting objectives. In: Stahl RG, Bachman R, Barton AL, Clark JR, deFur PL, Ells SJ, Pittinger CA, Slimak MW, Wentsel RS, editors, Risk management: ecological risk-based decision making. Pensacola (FL): SETAC Press, p 41–56.

Belzer R. 2001. Using economic principles for ecological risk management. In: Stahl RG, Bachman R, Barton AL, Clark JR, deFur PL, Ells SJ, Pittinger CA, Slimak MW, Wentsel RS, editors. Risk management: ecological risk-based decision making. Pensacola (FL): SETAC Press, p 75–90.

Bingham G, Bishop R, Brody MS, Bromley D, Clark E, Cooper W, Costanza R, Hale T, Hayden G, Kellert S, Norgaard R, Norton B, Payne J, Russell C, Suter GW. 1995. Issues in ecosystem valuation: improving information for decision making. Ecol Econ 14:73–90.

Campbell DE. 2001. Proposal for including what is valuable to ecosystems in environmental assessments. Env Sci Technol 35: 2867–2873.

Di Guilio R, Benson W, editors. 2002. Interconnections between human health and ecological integrity. Pensacola (FL): SETAC Press. 110 p.

Eisner T, Lubchenco J, Wilson EO, Wilcove DS, Bean MJ. 1995. Building a scientifically sound policy for protecting endangered species. Science 268:1231–1232.

McDaniels TL. 1995. Using judgment in resource management: a multiple objective analysis of a fisheries management decision. Oper Res 43:415–426.

Miranda ML, Mohai P, Bus J, Charnley G, Dorward-King E, Foster P, Leckie J, Munns WR Jr. 2002. Interconnections between human health and ecological integrity: policy concepts and applications. In: Di Guilio R, Benson W, editors, Interconnections between human health and ecological integrity. Pensacola (FL): SETAC Press, p 15–41.

Nash S. 1991. What price nature? BioScience 41:677–680.

[NRC] National Research Council. 1999. Perspectives on biodiversity: valuing its role in an ever-changing world. http://books.nap.edu/books/030906581X/html/index.html. Accessed May 5, 2007.

[NRC] National Research Council. 2004. Valuing ecosystem services: toward better environmental decision-making. Committee on Assessing and Valuing the Services of Aquatic and Related Terrestrial Ecosystems. Washington (DC): National Academies Press. 290 p.

[SETAC] Society of Environmental Toxicology and Chemistry. 2003. Valuation of ecological resources in environmental decision making workshop. October 7–9, 2003. Pensacola, FL.

Stahl RG, Bachman R, Barton AL, Clark JR, deFur PL, Ells SJ, Pittinger CA, Slimak MW, Wentsel RS, editors. 2001. Risk management: ecological risk-based decision making. Pensacola (FL): SETAC Press. 192 p.

[USEPA] US Environmental Protection Agency. 1997. Priorities for ecological protection: an initial list and discussion document for EPA. EPA/600/S-97-002. Washington (DC): USEPA, Office of Research and Development.

[USEPA] US Environmental Protection Agency. 2000. Regulatory economic analysis at the EPA. Washington (DC): USEPA, Office of Policy and Reinvention, Office of Economy and Environment. http://yosemite.epa.gov/ee/epa/eed.nsf/webpages/BrowsableReports.html. Accessed March 15, 2005.

[USEPA] US Environmental Protection Agency. 2001a. Improved science-based environmental stakeholder processes. EPA-SAB-EC-COM-01-006. Washington (DC): USEPA, Science Advisory Board.

[USEPA] US Environmental Protection Agency. 2001b. Understanding public values and attitudes related to ecological risk management: an SAB workshop report of an EPA/SAB workshop. EPA-SAB-EC-WKSP-01-001. Washington (DC): USEPA, Science Advisory Board.

2 Sociocultural Valuation of Ecological Resources

*Jessica Glicken Turnley, Michael D. Kaplowitz,
Orie L. Loucks, Beth L. McGee, and Thomas Dietz*

CONTENTS

2.1	Introduction	10
2.2	What Is Value?	10
2.3	Whose Values Are of Interest?	12
	2.3.1 Values of Individuals and Communities	14
	2.3.2 The Importance of Context	14
2.4	How the Concept of Sociocultural Value Relates to Economic Value	16
2.5	Why Are Sociocultural Values Placed on Ecological Resources?	16
2.6	Ecological Values and Public Policy	17
	2.6.1 Public Sector Decision-Making Mechanisms	19
	2.6.2 Natural Resource Management Successes and Failures	19
2.7	Valuation Approaches in the Private Sector	20
	2.7.1 The Business Impacts Valuation Problem	21
	2.7.2 Examples of Private Sector Impact Valuation and Decision Making	21
	2.7.3 Prospect	26
2.8	Data Collection	26
	2.8.1 Data Collection Processes	30
	2.8.2 Qualitative and Quantitative Data	31
	2.8.2.1 Methods for Eliciting Qualitative Data	32
	2.8.2.2 Methods for Eliciting Quantitative Data	33
	2.8.2.3 Value of Multiple Methods	33
2.9	Conclusion	33
References		34
Appendix 2.1		40
A.1	Data Collection Processes	40
	A.1.1 Methods for Collecting Qualitative Data	40
	A.1.1.1 Focus Groups	40
	A.1.1.2 Individual Interviews	42
	A.1.2 Methods for Collecting Quantitative Data	43
	A.1.2.1 Attitude Surveys	43
	A.1.2.2 Valuation Surveys	43

2.1 INTRODUCTION

The title of this book, *Valuation of Ecological Resources*, begs the question of the definition of "value" and, consequently, what it means to value a resource. This, in turn, raises questions of who is doing the valuing, on whose behalf, and for what purpose, and how values (once ascertained) will influence decision-making processes.

This chapter sets the stage for the more detailed expositions of methods and approaches appearing later in the book. In this chapter, we address the question of how to define value, begin to document how to identify the locus of relevant values, and reemphasize the rationale for these concerns in both the public and private sectors. We discuss values in the context of individuals, groups, and communities and strive to shed light on the differences in how each entity views valuation. We also outline data elicitation methods (see Appendix 2.1 at the end of this chapter) that underlie later discussions of analytic techniques and decision-making processes. Note that this chapter, as well as most of the book, assumes a Western, rational democratic (or at least pluralistic) approach to decision making.

2.2 WHAT IS VALUE?

There are 2 somewhat distinct traditions for the use of the term "value" in the social sciences, and both are relevant to the problem of environmental decision making. One is based on social psychology. The other is based on environmental and natural resource economics. The value tradition resting on social psychology focuses on the relatively unchanging priorities that individuals assign to actions and to the state of the world. In this case, "value" can be generally defined as "the principles or moral standards of a person or social groups, the generally accepted or personally held judgment of what is valuable and important in life" (Brown 1993). Public discussion of values is typically about these priorities — what kinds of things people consider to be important. With regard to environmental decisions, several kinds of these social psychological values have been investigated.[1] These include the following:

- Self-interest: a desire for power, authority, and wealth.
- Traditional: a concern with order, respect for elders and the past, and a focus on the family.
- Openness to change: a desire for new experiences and innovation.
- Altruism: a concern with the well-being of others. Altruism can in turn be subdivided into 2 components: "humanistic altruism," which is a concern with the well-being of other humans, and "biospheric altruism," which is a concern with the well-being of other species or the biosphere itself.[2]

---

[1] The literature on the social psychology of values flows from Kluckholn (1952) through Rokeach (1968, 1973) to Schwartz (Schwartz 1987, 1992; Schwartz and Bilsky 1990). The application of this concept to environmental concern was first introduced by Dunlap et al. (1983) and further developed in a series of more recent papers (Stern et al. 1993, 1999; Stern and Dietz 1994; Stern 2000).

[2] While it is sometimes argued that those who care about the environment don't care about human welfare, the empirical evidence for the general population of the United States refutes this claim. Among the US public, there is a very strong positive correlation between concern with other humans and concern with the environment (Stern et al. 1999; Dietz et al. 2002).

Independent of this literature, a robust tradition of estimating the social value of environmental goods and services has emerged in resource and environmental economics. This tradition, based on econometric examinations of the trade-offs people are willing to make concerning environmental goods and services, will be reviewed in detail in later chapters of this book. It should suffice to say at this point that estimates of individual and aggregate social values associated with environmental and natural resources historically have been the purview of economists. Here our main concern is to illustrate the relationship between individual values as described in the social psychological literature, and values as used in day-to-day discussions and as measured based on a social sense of value.

There is an additional important dimension of value that intersects the personal and social aspects of value. This dimension rests upon the distinction between "enduring values" and "preferences." Enduring values are generally thought to be fundamental notions associated with a good or service, and are assumed to be independent of any (social) time or place and to be a property of the thing itself. Preferences are posited to be relatively ephemeral, to be time and space specific, and to be statements of attitudes of individuals. Preferences may reflect enduring values to a greater or lesser degree. Note, however, that it can be argued that even notions of the "good" are time and place specific. The distinctions between these 2 philosophical approaches are presented more fully in Chapters 3 and 4. Here we note that the distinction between enduring values and preferences can affect not only the ultimate value placed upon resources but also the process by which the valuation is made.

In addition to the various explanatory traditions regarding values, there are multiple intellectual traditions regarding how to make decisions about activities that impact the environment. While the logical bases for these approaches have been developed within ethical and political theory, they present themselves in public activity. So in addition to value differences, a manager may face differences with regard to how people think decisions should be made (concerns about process), not just about what the decision should be (concerns about outcome).

Within environmental policy, the utilitarian stance of the greatest good to the greatest number has been the most extensively and robustly developed decision-making approach. Note the assumption that the greatest good to the greatest number of individuals equates to the greatest social good (the social good is an aggregate function). We will return to this point later, but it is important because the utilitarian approach is the fundamental ethical logic that underpins economic valuation of environmental goods and services and the use of benefit–cost analysis to make trade-offs across outcomes. The core idea is that one attempts to estimate the benefits and costs that will be visited to every individual in society for each decision option. Then one sums the benefits and costs across all individuals to assess overall social benefits and costs, and chooses the option that provides the greatest overall benefits relative to costs.[3] There are, however, several other approaches in ethical theory that seem

---

[3] This approach has been criticized because it ignores the distribution of costs and benefits across individuals and groups. In practice, this is usually the case. But in principle, benefits and costs to some groups could be given greater weight than those to others. However, one then has to find a way to decide what weights to assign, and there is no consensus on this issue (Freeman 1993).

to make sense to many members of the public but have been less well developed in policy analysis.[4]

Kantian ethics accepts utilitarianism within bounds, but notes that there are some states of the world, and some decisions, that are inherently unacceptable, whatever the changes in human welfare they generate. In some cases, this position rejects the use of economic valuation; in others, it argues for "guardrails," "safe minimum standards," or "precautionary principles." Such limits on strict utilitarianism allow for implementing and addressing such concerns as "out-of-bounds" actions or states of the world. This Kantian approach forms the basis of such policies as the Endangered Species Act (ESA, passed in 1973) that explicitly states that economic analyses may not be the basis for listing (and protecting) species. In these types of cases, there is a "bright line" developed according to nonsocial criteria (in the case of the Endangered Species Act, ecological criteria are used), and actions or events that fall below that line are proscribed. The sociocultural value is reflected in the judgment to exclude economic analyses from the decision-making equation on the presumption that species are important in and of themselves, that is, have an intrinsic "eternal" value. Such "intrinsic" values begin to approach the notion of enduring values introduced earlier.

Deliberative ethics is also a major contender for a place in environmental policy analysis. The deliberative approach derives from the democratic theory of John Dewey (Dewey 1923) and Jürgen Habermas (Habermas 1970, 1993) and has been proposed for environmental policy analysis by many (Dietz 1987; Fiorino 1990; Webler 1993; Dryzek 1994; Renn et al. 1995; Jaeger et al. 2001; Rosa et al. 2000). The core idea of the deliberative approach is that good policy emerges from a dialogue among all parties who will be affected by a decision. Furthermore, such dialogue must be fair in the sense that everyone is on an equal footing and must be competent in the sense that the best science is brought to bear in the discussion. Several recent policy reports have further developed and endorsed the idea of the deliberative approach for environmental decision making (National Environmental Justice Advisory Committee 1996; Stern and Fineberg 1996; US National Research Council 1999). These notions are also codified in such policy statements as the 1994 Executive Order (EO) 12898, "Federal Actions to Address Environmental Justice in Minority Populations and Low-Income Populations."

2.3 WHOSE VALUES ARE OF INTEREST?

The preceding exposition of the notion of value and its relationship to ecological resources emphasizes that the conversation does not treat those resources in a vacuum but rather in relation to some human activity. Human activities can be described in terms of 3 elements: individuals, networks of social relationships

---

[4] These approaches are reviewed in US National Research Council (1999). The contrast between utilitarian and deliberative ethics is described in Dietz (1994), Dietz and Stern (1998), and Jaeger et al. (2001).

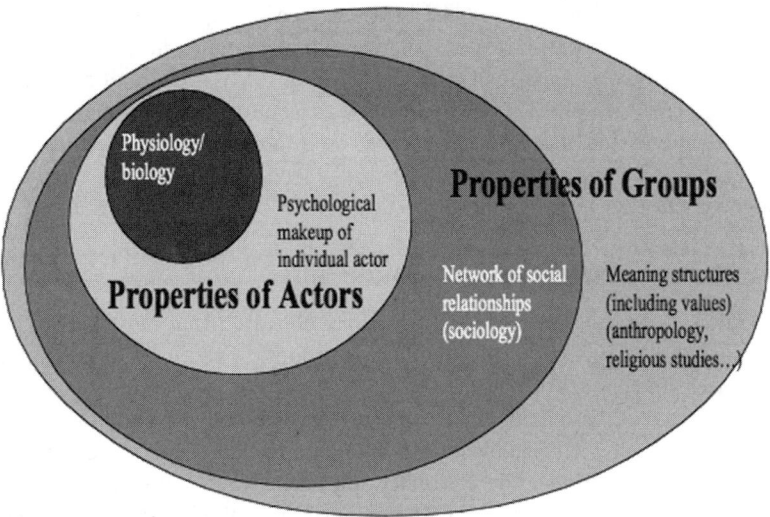

FIGURE 2.1 The context of behavior after Spulak and Turnley (2005).

generally organized into institutions,[5] and the constellation of symbols and meanings (including values) that groups construct to make sense of the world and provide a structure for individual choice and action.

These concepts may be recast (Figure 2.1) to posit the action of any individual is the function of the intersection of several factors:

1. the physiology of the individual;
2. his or her psychology (which itself is a combination of physiological parameters and individual life experience, including the culture within which a person lives);
3. his or her position in a web of social relations, themselves organized into institutions and organizations of varying levels of formality and having a life beyond the moment of his participation; and
4. the value, or importance, the group places upon a target action, that is, the group's understanding of the positive or negative consequences of (non)performance of that action within both symbolic and "real" worlds.

It is important to emphasize that this is not a reductionist approach (i.e., if we just know enough about physiology we could explain the world; see Wilson 1975, 2000; Richerson 1977). It also is not an additive approach. It is, rather, a dynamic systems approach in which each element is partially explained in terms of the others, all of which are constantly changing over time. In the context of this book, an individual's values are partially a reflection of his or her own individual history, and partially

---

[5] Parsons (1949) defines institutions as "the body of rules governing action in pursuit of immediate ends in so far as they exercise moral authority derivable from a common value system." They thus are phenomena of groups, endure over time, and are invested with an authority other than force that can stimulate compliance. They may or may not be formalized as "legal" or corporate entities.

a reflection of his or her sociocultural location. In a sense, then, each action "says" something about both the individual and the world in which he or she is situated. For example, a Native American growing up on a fairly traditional reservation in the American West might view the bald eagle as a source of power and strength for the individual and treat it accordingly. However, if the same individual was raised in Albuquerque or San Francisco, he or she might see the same bird as a symbol of a nation-state and engage in a completely different set of actions toward it.

2.3.1 Values of Individuals and Communities

Despite the systems nature of the explanatory construct that must be engaged in order to understand the behavior of humans in groups, institutional, methodological, and historical factors have combined to disaggregate the whole into separate fields of analysis. Each of these fields or elements has its own body of methods for analysis and is generally characterized by certain methodological and associated ontological or constitutive assumptions. The 2 primary approaches are methodological individualism, which takes the individual as the primary unit of analysis, and a collective approach, which begins with a group.

Methodological individualism suggests that "all social phenomena can only be explained in terms of individuals and their interactions" (Udehn 2002). The ontological underpinning of this approach is that only individuals are real and (in our context here) only the values of individuals can be measured. Values of groups are defined as the aggregation of the values of the individuals who compose them, that is, functions of, and only of, the actions of the constituent individuals. Countering this approach is the notion represented by Émile Durkheim's concept of "collective consciousness," which he defines by saying that "society is not a mere sum of individuals. Rather the system formed by their association represents a specific reality which has its own characteristics." He goes on to say that "nothing collective can be produced if individual consciousnesses are not assumed; but this necessary condition is by itself sufficient" (Durkheim 1964).[6] Values in this context may be seen (in the language of complexity science) as emergent

Sociocultural value thus has several characteristics. It is manifested or expressed through individual activity, although it is partially a function of dimensions that existed before and after any particular individual is present (phenomena of the group: social structure and cultural complexes). However, because it is partially a function of the particular history of the individual expressing it, no 2 individuals will express the "same" value in the "same" way.

2.3.2 The Importance of Context

If the definition of "value" depends to some degree on the sociocultural location of the individuals who express it, we are begging the question of community and how

---

[6] This tension between the individual and the collectivity in scientific explanation will be familiar to those trained in evolutionary biology, where Darwinian "population thinking" acknowledges the importance of the individual but also understands that the collectivity of individuals forms the environment for the individual (Mayr 1959; Sober 1980; McLaughlin 2001).

that is defined in the decision-making process. It is particularly important to keep in mind that, like "stakeholder" or "interested party," "community" in this instance is defined relative to a particular issue or action (see Glicken 2000). This notion of community has both spatial and temporal dimensions. Although this aspect of evaluative complexity will be addressed later, a few points are worth mentioning here.

Communities and their associated values change over time. There are many theories about how this change takes place. These range from rational actor approaches which seek to explain the emergence of communal structures in a world of self-interested beings (Hobbes [1651] 1998; Axelrod 1984), to ethnomethodology and transaction analyses which search for understanding of how individuals interact to create "scenes" or "scenarios" (Goffman 1959, 1974; Garfinkel 2002). These theories of change are important for the decision maker only if he or she seeks to engineer value shifts deliberately.

It is important for the decision maker to recognize that a community's current values are not necessarily those it will espouse in the future. A good example of how community values may change over time is embodied in the story of attitudes toward large-scale dams. Once seen primarily as economic engines (the creation of jobs and wealth through both their construction and their subsequent economic contributions through electricity generation and enhanced agriculture) and as positive contributors to improved quality of life through rural electrification programs, ascendant concerns about ecological damage have cancelled the construction of some planned large-scale dams around the world, and led to the planned destruction of others, particularly in the United States. Thus the decision maker must factor into his or her decision concerns about future communities' values and generational equity. We will return to this later. In addition, the consequences (success or failure) of certain types of actions related to the environment, such as restoration of certain ecosystems, may not be realized within a human lifetime.

Certain types of communities, such as political jurisdictions, are spatially defined. The boundaries of such systems may not be congruent with ecological systems (watersheds crossing many local and regional geopolitical boundaries are excellent examples). Furthermore, some ecological features may not have value for a community local to it, but may contribute significantly to a broader ecological cycle. The case studies of the Rio Tinto mine in Madagascar and the oil spill in the Arabian Gulf after the 1991 Gulf War presented at the end of this book (Case Studies 3 and 4, respectively) are examples of the "nested" nature of communities and their associated value structures (see also Gibson et al. 2000). Whose values "count" in these cases? Is it appropriate to override local concerns for those of a large community? Who makes that decision? And who bears the associated responsibility for maintaining the services provided by the target ecosystem? How do we reconcile this with existing political realities? These kinds of questions will be addressed later in this section, as well as throughout the book.

The notion of valuation of ecological resources thus begs several questions, which can only be answered contextually. As a corollary, concerns raised by the relative nature of the definition of community have significant implications for the collection of data on values. As communities must be defined anew for each discussion, so must the data on values be collected each time. This "curse of specificity" can have an impact on project planning and data analysis as elicitation of this type of data can be very resource intensive.

2.4 HOW THE CONCEPT OF SOCIOCULTURAL VALUE RELATES TO ECONOMIC VALUE

The economic sense of value rests on the "willingness to pay" of individuals. It is assumed that an individual's willingness to pay for a good or service reflects the preference she or he has for that good or service, given other ways in which individuals might allocate their funds. Again, we leave the details of this approach to subsequent chapters. But here we note that a willingness to pay for an environmental good or service can be thought of as a behavioral manifestation of more general and enduring values, including a concern with the state of the environment per se or of other species (biospheric altruism), the welfare of other people (humanistic altruism) or one's self-interest, which may include both perceived prevention of environmental harm to oneself (a benefit) and economic losses to oneself because of environmental regulation (a cost).[7]

Put simply, concern with the environment as expressed in willingness to pay is a reflection of more general and relatively stable values. Individuals differ in those values, and some of these differences may be related to social structure and culture, including gender and ethnicity (Dietz et al. 2002; Kalof et al. 2002), though these relations have not been extensively investigated. As suggested earlier, a determination of an individual's sense of the economic value of some ecological resource can be contextualized and more completely understood through an analysis of community-based sociocultural value (Boxall and Adamowicz 1999). An economic, monetized, or quantified expression of value is thus recast as the behavioral manifestation of a wide variety of motivators or drivers for that behavior. Gaining a deeper understanding of the motivators is very important for a decision maker who wishes to understand either what it will take to change behaviors or the types of secondary effects a particular decision on his or her part might engender.

Economic analyses generally are directed toward the achievement of efficiency. To follow our statement above, they are designed to determine the minimum amount necessary to "pay" (give up) to get the desired object. Particularly when conducted at a group level (i.e., to determine the minimum amount the community as an aggregate of individuals is willing to pay), they do not address questions of distributional fairness or equity. As we noted earlier, deliberative approaches to decision making can highlight such inequities and allow for greater distributional fairness either through the moral imperatives that emerge from the deliberative process itself, or from the imposition of codified imperatives such as EO 12898 (the environmental justice executive order).

2.5 WHY ARE SOCIOCULTURAL VALUES PLACED ON ECOLOGICAL RESOURCES?

In the Western philosophic construct, human communities are increasingly seen as an integral part of the environment. Over the last several decades, contributions by thinkers such as Aldo Leopold (1949) and Rachel Carson (1962) introduced the notion of stewardship, suggesting that the natural world is not limitless and that

---

[7] Research on the correlates of public environmental concern also shows that those holding traditional values are often opposed to environmental regulation (Stern et al. 1999; Whitfield et al. 1999).

Sociocultural Valuation of Ecological Resources

human communities may induce irreversible changes that will affect the sustainability of the environment on which we depend. This view of environmental stewardship reached some of its most extreme statements in the deep ecology movement (Naess 1972 in Miller 1991; Devall 1986), but all were part of a trend that shifted conceptions of the environment from a source of available, unlimited resources and associated ethics of exploitation, to one of finite resources and an associated ethos of stewardship and responsibility (Nash 1967; Glicken and Fairbrother 1998). The related notion of embeddedness rather than mastery over and separateness from the environment was epitomized by Lovelock's (1979) Gaia concept, which emphasized the interconnectedness of all natural systems and our responsibility (as one of the systems' parts) for the health of the whole.

2.6 ECOLOGICAL VALUES AND PUBLIC POLICY

This change in perception of our relationship to the environment can be viewed through the lens of public policy. The establishment of the US Environmental Protection Agency (USEPA) in 1970 was a collective statement that public resources should be devoted to protecting the environment. The USEPA's own description of its history reads,

> In July of 1970, the White House and Congress worked together to establish the EPA in response to the growing public demand for cleaner water, air and land. Prior to the establishment of the EPA, the federal government was not structured to make a coordinated attack on the pollutants that harm human health and degrade the environment.[8]

The Clean Water Act (as amended in 1977) mandates that we "restore and maintain the chemical, physical and biological integrity of the Nation's waters" recognizes our society's value for aquatic resources. Similarly, the Endangered Species Act can be viewed as a manifestation of our value for preserving and conserving biological diversity. The role of the government, in this context, is to ensure that these valued public resources are protected. The role of government in responding to these ecological resource issues changes over time, reflecting, in part, changes in societal values for these resources.

Environmental decision making, as noted in Chapter 1, can be viewed as a hierarchical process, ranging from policy making by senior government officials to technical and logistical decisions made by engineers, technicians, and scientists. Our emphasis is on risk managers who are responsible for implementing and enforcing environmental policies and laws. In this context, the risk manager operates primarily in a reactive mode — to enforce regulations or respond to some emergent need or existing problem. The kinds of decisions commonly made by public environmental risk managers include the following:

- Setting standards to protect human and ecological health
- Registration of new chemicals and products

---

[8] http://epa.gov/epahome/aboutEPA.htm.. Accessed May 4, 2007.

- Major construction projects
- Site remediation
- Natural resource management

Public environmental risk managers are often faced with having to balance competing public demands driven, in part, by different societal values for a particular natural resource. For example, the CALFED Bay–Delta Program (see Case Study 6 at the end of the book) was initiated primarily in response to competition for water for agricultural irrigation and urban use versus water needed to support migrating Chinook salmon, a federally endangered species. Initially, an "all or none" approach was adopted, whereby winter-run salmon were provided the necessary water for migration, regardless of the economic hardship to agricultural and urban water users. As the salmon population rebounded, a compromise was reached. Funding for "ecosystem restoration" and to purchase "environmental water" that could be managed by the natural resources agencies was given in exchange for the assurance of reliable water for agriculture and urban uses; however, questions relative to valuation remain. For example, are the amounts of water and money enough to balance the potential adverse effects and lead to restoration of the species? How should the money for restoration be allocated among the multitude of restoration projects? The CALFED Program also illustrates the importance that the affiliation of the decision maker (e.g., federal, state, local, or tribal government) has on the valuation of natural resources. Differences also may occur both among agencies that have similar statutory mandates but have responsibility for managing resources at different spatial scales (e.g., the USEPA, and local municipalities and state agencies responsible for water quality regulation) and across agencies with different statutory mandates (e.g., the USEPA, US Fish and Wildlife Service, and US Bureau of Reclamation).

Another difficulty in incorporating sociocultural values into environmental decision making is the lack of congruence between boundaries of political jurisdictions and ecological systems. An example is the management of blue crab populations in Chesapeake Bay. The blue crab can be considered the symbol of Chesapeake Bay and is engrained in the culture and livelihood of many bay area residents. Crab population decline appears to be related to fishing pressure and water quality. Like many commercial species, the life cycle of the blue crab knows no political boundaries, traveling through Maryland and Virginia waters during different stages of development. Efforts to manage this resource on a baywide scale, however, have been largely unsuccessful due to the inability to satisfy competing interests (Ernst 2003). Each state has an economic stake from the commercial and recreational value of the fishery, as well as the packing and processing of crabmeat. Crab regulations designed to manage this fishery in a sustainable way may have a disproportionate effect on the economy of the states (the issue of distributional fairness, mentioned earlier); furthermore, within each state, a small number of areas that rely heavily on the crabbing industry also may be disproportionately affected. For example, in communities such as Tangier Island (Virginia), Smith Island (Maryland), and Crisfield (Maryland), commercial fishing is the main source of income. Crabbing restrictions are particularly hard felt in these communities as they are viewed as a threat not only to the livelihood of the commercial fishermen, known regionally as "waterman," but also to a way of life that goes back many generations. The rich culture of the watermen communities is part of the Chesapeake Bay heritage, and there is strong interest in preserving it. The challenge for natural resource managers is to reconcile the economic needs of the states, the value of preserving the culture of the Chesapeake Bay watermen, and the sustainable management of the blue crab population. Environmental risk managers need tools that can be used to compare and evaluate these competing values as well as consider the interests and values of environmental groups and industry.

2.6.1 PUBLIC SECTOR DECISION-MAKING MECHANISMS

Mechanisms that are used to balance competing private and public interests for a particular natural resource include regulations, incentives, and management structure. Public environmental decision making includes various combinations of these approaches.

Traditionally, regulations have formed the basic foundation for environmental protection policies. Regulations dictate, to varying degrees, the level of protection that is to be afforded to natural resources. For example, the Endangered Species Act is far more prescriptive in the protection of endangered species than the Clean Water Act is for the protection and enhancement of water quality. As noted earlier, the ESA has its basis in Kantian ethics where a "bright line" is established (based on nonsocial criteria) and actions that cause species to fall below that line are proscribed. Similarly, in the CALFED case study, the "all or none" decision to allocate water for the Chinook salmon when numbers were extremely low also exemplifies this theory. The level of environmental protection in many cases is not prescribed, leaving the regulator the task of integrating economic, sociocultural, and ecological values into the decision-making process. In addition, as noted above, in many instances regulation is ineffective because the resource or environmental issue is not being managed in a holistic way.

Economic incentives are playing an increasingly important role in pollution control (Pearce and Turner 1990) and provide an instrument to achieve protection of ecological resources. Examples include emissions trading applied in the US air pollution context and subsidies such as tax breaks for the incorporation of a designated type of pollution technology. One much praised approach is the use of Tradable Environmental Allowances (TEAs) in which the right to emit pollutants or use a resource is traded in a market (Rose 2002; Tietenberg 2002). This encourages efficiency in dealing with environmental problems and provides a price for the value of the resources. Care must be taken, however, in designing such markets to match the characteristics of the resources and the users.

Finally, management structures, such as the CALFED Program, allow for the discussion and consideration of competing interests and values, with the goal of determining solutions that consider these values in an equitable way. This program is an example of applying the "deliberative approach," described earlier, to derive environmental policy where all affected parties are involved in the decision-making process. The success of these types of programs may, however, in some instances, be dependent on whether or not they are endowed with regulatory authority (Ernst 2003).

2.6.2 NATURAL RESOURCE MANAGEMENT SUCCESSES AND FAILURES

The examples below illustrate the influence that sociocultural values can have on the success or failure of natural resource management.

In 1978 the bald eagle was listed throughout the majority of the lower 48 states as "endangered" on the federal list of threatened and endangered species. As indicated previously, the ESA can be viewed as the legal embodiment of our nation's values for biological diversity and species protection. The cause of population decline was related to hunting and to reproductive failure caused by the accumulation of DDT (dichlorodiphenyltrichloroethane). In 1995, the bald eagle was reclassified as "threatened" in the lower 48 states. The recovery of the bald eagle is a success story for the ESA and the management and protection of a valued public resource. Its importance to the citizens of the United States clearly goes beyond its "ecological value" as an endangered species; it is a national symbol and a reflection of pride in the United States. It seems likely that management efforts were successful, in part, because this was such a universally valued resource. Is the protection of less "charismatic" endangered species, such as the snail darter, more difficult because they are "valued" less by a society as a whole?

Lack of consensus or disparate views of the value of natural resources can result in seemingly irresolvable conflicts. The issues surrounding the allocation of water in many western states arise from the intersection of many different value sets. Public good versus private interests is one. Individuals, not the state, hold water rights ... yet in many cases, water allocation decisions pit the public good (e.g., drinking water) against individual quality of life (e.g., residential use). The conflict between different use sectors is often the most public area of disagreement over water. Is agricultural use more important than industrial use? More important for whom? And valued by what scale? The CALFED case study illustrates an interesting way in which environmental values were brought into play by establishing a new "use category" for water ("environmental use"), which allowed environmental values legal status equal to that embodied in other use categories such as industrial, agricultural, and residential.

2.7 VALUATION APPROACHES IN THE PRIVATE SECTOR

In addition to the broad public concern for environmental and sociocultural values described above, the private sector has had to consider them. These considerations have often been local, in the vicinity of new projects or facilities, but they can address large-scale risks to air, water, or the consumption of resources that a community values. Sometimes a multinational company must make worldwide public judgments about impacts of its operations anywhere in the world. The public's concern may be in relation to perceptions that local resources are being put at risk, that globally significant biological diversity is being put at risk, or that local to regional culture and values are put at risk. Major projects associated with some extractive industries can induce local human migrations that bring still other negative consequences for the environment and social stability.

The private sector needs to be concerned about these environmental and cultural value outcomes because, in the aggregate, they can lead to negative effects on a firm's product acceptability, market share, profitability, and potentially shareholder value (Ditz et al. 1995; Erekson et al. 2000). Many companies have adopted environmental management systems that require "continuous improvement" in their corporate environmental and sociocultural performance (Erekson et al. 2000). Thus, for reasons rooted in processes similar to those seen in the public sector, a new standard is emerging by which business and industry are considering local to global expressions of cultural values and are incorporating mitigative, management, or impact avoidance measures into new project decision making and routine operations.

2.7.1 THE BUSINESS IMPACTS VALUATION PROBLEM

From some viewpoints, industry responses to environmental or social impact problems may be seen as simply achieving compliance with regulations established through the public sector. However, the standard that is emerging is going "beyond compliance," as is treated comprehensively by Smart (1992). Compliance with regulations can be seen as a minimum response, one that does not move a firm toward opportunities for competitive advantage (and potential market share), which accompanies distinguished corporate citizenship and role modeling. In addition, many sociocultural elements are not covered by regulations in any country, including many new expressions of sociocultural value (as is seen in the conservation of unique indigenous genetic stocks). Recently, there have been many initiatives by industries and trade associations to develop standards for their members that go beyond the requirements of current regulations, motivated by both an interest in protecting the environment and a hope of avoiding further regulation. In some cases, government encourages and facilitates these programs. However, little is known about their actual effectiveness in reducing adverse environmental effects (Dietz and Stern 2002).

At the same time, companies operating existing facilities must respond to community concerns about technical and sociocultural "disruptors" (such as odors or "burps" at some petrochemical facilities). The costs of controlling these must be judged in relation to the value the community attaches to their avoidance. However, the company can also translate whatever measures it takes into a category of lower maintenance costs and more efficient site-specific business practices, generally. Under optimal conditions, achieving congruence between the costs of minimal impact operations and the environmental values of the community or region could add value to companies and their shareholders (Erekson 2000).

These practices have sometimes been referred to as the "greening of business," which includes waste minimization and efficient multiple use of resources and land. Nearly all of these measures have the potential to reduce operating costs and improve profits. Environmental elements of business planning are now being implemented by a few corporations. They may be establishing a new standard of environmental performance. The new standard, if it is accepted, will go beyond what has been known as greening, and will consider effects on sociocultural values as well as ecological values. Some of these companies will be building an information base that they can act on for reasons of their own self-interests, including competitive advantage.

2.7.2 EXAMPLES OF PRIVATE SECTOR IMPACT VALUATION AND DECISION MAKING

As a first step, it may be useful to consider a sociocultural valuation case study undertaken by a private, not-for-profit organization, The Nature Conservancy (TNC), in the early 1980s (Leavitt 1994). The site is the then emerging Virginia

Coast Reserve, a complex of 13 barrier islands stretching 80 miles along the Atlantic coast of Virginia.

Coastal bird populations on the islands depend importantly on the health of the intercoastal lagoons, which in turn is determined by land use and runoff discharges from the adjacent uplands. These mainland properties were all potentially subject to intensive shoreline development that, to the north in Maryland, had led to degradation of lagoon systems and their biological diversity. To make a decision on its conservation strategy, TNC undertook intensive surveys of the sociocultural and environmental values of local communities along the coast of a 2-county area. A key component of the survey sought to determine the depth and scope of conservation values in the local community, and whether the community welcomed the prospect of potential wealth from selling land to the out-of-state developers and working in condominium construction.

Somewhat surprisingly, the surveys demonstrated a strong conservation ethic among the local residents, but also a reluctant resignation as to development of shoreline condominiums because they saw no means of a sustainable livelihood through conservation alone. As a result of the demonstrated conservation values, TNC made a decision to assist the community in grassroots for-profit ecotourism, marketing of unique local produce, and low-impact, low-density conservation summer homes (Leavitt 1994). These were the first steps by TNC toward the "community-based conservation" that is now a hallmark of their conservation strategy worldwide. It should be noted that the data collection and analysis were more narrowly prescribed than are usually done for public sector policy positions or decisions, but were unconstrained by the requirement of a set return on investment that characterizes the private sector examples that follow.

Most for-profit industries, however, face serious difficulties in assessing public concerns due to the wide range of impacts that affect their businesses (local to global) and their limited expertise in such work. A simple example is evident in the case study on the Fetzer Wine Company's decision to gradually end the use of pesticides in its vineyards (Erekson and Leavitt 1999).

Fetzer Wine Company was aware of the public's perception of risks to public health and biological diversity from its use of many insect and disease control agents, but assumed it had no alternative. However, it maintained a large visitors' reception garden, free of pesticides, in which it grew high-quality fresh produce along with grapes. The grapes were found to be more flavorful than grapes grown nearby on pesticide-treated fields. After careful consideration of its own corporate culture and values, along with values held by its customers and the community, Fetzer decided to end nearly all uses of biocides and allow the soil and soil fauna to recover. The costs of manual control of insect and disease problems seemed likely to be greater than past practices (despite savings in pesticides and fertilizer), but the sociocultural values they espoused appeared likely to be worth the risk. The company did no survey of values held by neighbors or the local community, but did perform an analysis of consumer trends toward premium organically grown food products. In the end, the company's decision to end pesticide use was also influenced by the potential for an upscale new product, the organically grown Fetzer Bonterra wine. With good management, the company had an opportunity to create shareholder value at the same time as production systems were changed, and ecological values restored.

Another example illustrates how a multinational corporation may need to be concerned with values held by the global community, along with conflicting values held locally around major projects or facilities. The development of a new ilmenite mine from sand deposits in Madagascar (for titanium) by the multinational company Rio Tinto is an example (see Case Study 3 at the end of the book). The company wanted to be sensitive to the worldwide values associated with unique biological diversity on the development sites in Madagascar. At the same time, some of the local Malagasy people wanted to see the development proceed for poverty alleviation, without immediate regard to biodiversity risks (although long-term or irreversible outcomes were a concern). Others in the local community valued the sustainability of the wood-fuel harvest from the mine site and sought reassurance that forest regeneration would be achieved during and after the mining operation. Their goal was to maintain the integrity of their lifestyle, social structure, and language. Quantifying these multiple expressions of value was challenging enough, but resolving which array of values ultimately might prevail (the global versus the local) required a complex analytic process. The company's Environmental Plan (see Case Study 3) has tried to make balanced choices, having incorporated a substantial component of education for all stakeholders. Still, the methods used are not yet standardized or widely available.

Still another common industry valuation problem arises in the decommissioning of industrial or intense resource use sites. A depleted oil field is an example (see Box 2.1). Significant opportunities exist around oil fields for natural resource value creation, adding somewhat to local sociocultural values and opportunities. Certain expenses beyond essential cleanup are required on the part of the company, however, to create such value.

Box 2.1 Private Sector Land Management in Site Decommissioning

Decommissioning of operating sites presents important challenges as to the balancing of costs to corporations and benefits to ecological resources and communities. Not only does the reduction of operations result in lost streams of revenue for companies, but also mothballing, dismantling, and leaving the site in a safe and acceptable condition introduce new sources of operating expense. Although such expenses are finite, they can be significant. If an entity could identify sources of valuable extractable services from the facility, creating revenue or values that could offset the new operating expenses, the decision could be balanced and sustainable. Targets for cost reduction include site carrying costs (e.g., taxes, maintenance, and security), remedial costs (containment, removal, and destruction), and future (injury during reuse) liabilities. Additionally, new marketable ecological services could be created through wetlands mitigation or conservation banking measures. Extraction of the current or potential value associated with the site's ecological services is a possible avenue for adding shareholder value associated with property management. A key barrier to moving in such a strategic direction is the availability of effective tools to place a value on the ecological benefits created. It should be noted that, in contrast to the above model, utilities are currently required to recognize on their books and to fund in cash their estimated decommissioning and reclamation expenses throughout the active life of the associated plants. This might serve as a model for other industries.

Hypothetical Example

A 2000-acre site was used for the production of oil. Equipment and the operating plant footprint for the site are distributed mostly over the interior of the property. Along the eastern boundary of the site is a valuable and scarce riparian habitat. The water flow from this habitat flows to wetlands associated with a regional nature preserve on the southern boundary of the property. Additionally, in areas of the property that have not been disturbed, prairie pothole wetlands are found that provide important ecological services for local mammals, birds, and additional migratory populations.

Questions: can the values (financial, social, and ecological) of natural resource end uses or enhancements be factored into the site redevelopment plan? If so, what type of cost and benefit determinations would be used?

Facility components	Environmental footprint
1. Well heads	Environmental impacts include past spills of crude oil or brine, as well as impacted soils from access roads for maintenance of the wellheads.
2. Pipelines	Disturbed areas around pipeline right-of-ways.
3. Gathering stations	Disturbed areas along pipeline right-of-ways that come together, and crude oil or brine contamination associated with spills or pipeline leaks.
4. Degas(sing) plant and storage tanks	Physical disturbance from equipment siting, and incidental contamination associated with spills or equipment failure.
5. Gas plant waste treatment facilities	Physical disturbance from equipment siting, incidental contamination associated with spills or equipment failure, and residues of wastes in ponds or land farms.
6. Noncontaminated interior land*	Not directly impacted, but habitat may be fragmented and suboptimal; impacts can be ameliorated.
7. Buffer lands*	Not directly impacted, but habitat may affected by surrounding land use and interior land fragmentation; impacts can be ameliorated, and together with restored interior lands, a valuable habitat mass can be created.

* Total could add up to > 80% of land area.

In the site redevelopment plan, some of the following strategies for value extraction are of concern to the company, and others are of concern to the local or regional community:

- Facilitation of natural versus engineered habitats to lower maintenance costs
- Conveying a conservation easement to lower property taxes
- Property donation to lower income taxes

> - Ecological habitat enhancements to create potentially marketable ecological services (e.g., wetland mitigation banking)
> - Reducing the area of contaminated lands through facilitated biological treatments and restoration
> - Reducing corporate liability costs associated with natural resource damages
> - Enhancing company reputation value with agencies and the community
>
> The values listed here accrue to both the company and the community in different ways. All can be estimated, but at additional cost. Neither the company nor the community wishes to incur the cost of the value determinations so as to complete a benefits and costs assessment. Without the community benefits determination, the company is left with the cost of remediation well defined, but with little information on benefits. Unfortunately, a decision is likely to be made without the needed full information on both.

A problem arises from the relative absence of estimates for the value attached to the restored resource, in comparison to the detailed determination of expense being incurred by the company. While a company may accept the cost of habitat restoration in the interest of building its reputation for community service, its decision processes require some form of value determination originating with the community's interest in natural resource benefits. The cost of a survey to estimate such values is not generally an expense the company would want to incur, nor is the local public sector able to consider it. As discussed in Box 2-1, a decision will be made, but to the extent it is made with only limited information, it may not capture the full opportunity for benefits to both the community and the company.

Many large companies, however, have been able to allocate the resources needed for comprehensive assessments of where to undertake major new development. In this way, these companies are discharging their responsibility as sociocultural stewards, globally, regionally, and locally, at a cost that is seen as part of the new project and is consistent with the benefits in corporate reputation for responsible stewardship.

The new methods by which business and industry incorporate community-based values into decisions warrant more detailed consideration. Generally, businesses do not use a standard framework for problem formulation, data collection, or analysis, because the scientific requirements arising from each business's relationship to its community range so widely. In addition, companies tend to use activity-based environmental accounting methods to summarize their costs and potential benefits (Gorman 2000), rather than a broad economic or trade-off analysis. The examples above show that the private sector does, at times, collect data on community attitudes at various scales, from the siting of suburban filling stations to overseas pipeline development. However, much of the data on value are of a general form, unless the industry is concerned with potentially transforming projects such as described in the Rio Tinto case (see Case Study 3). The larger problem at this time is that most companies do not have the expertise to do such surveys. Further, they generally are not well informed as to how to do analysis and interpretation of data collected, or in using that analysis in reaching a decision on a project.

A few companies are becoming more sophisticated in their analyses, however, and have begun projecting sociocultural benefit transitions over the lives of major projects, often a period of 2 generations. These efforts are becoming a kind of cradle-to-the-grave approach to impacts and their resolution over time. An emerging standard involves analysis and projection of the future states of sociocultural attributes and values in local communities, domestically or overseas, beginning with full implementation and continuing through closure of the projects. These companies are considering all reasonable options for sustaining local to global sociocultural values. In the end, some projects may not proceed because the sociocultural costs cannot be managed, and other development options are available to the company. Such decision processes and outcomes are quite different from those seen in previous sections for the public and not-for-profit sectors.

Of course, a substantial problem remains. It is often the case that transforming an ecosystem may reduce the value of the ecosystem's services, especially its cultural services, while producing economic return in the market. The firm making the decision, and in some cases the local community, could be capturing the market-based returns from the ecosystem transformation. However, the value of cultural services such as the preservation of an endangered species or of biodiversity in general is a small per capita value for a very large number of individuals. The methods described here and detailed elsewhere in the volume allow us to estimate those values. However, they do not provide a mechanism for capturing those values in ways that allow them to be balanced against the market values that drive most private sector decisions. While this disjuncture is most true for ecosystem cultural services, it can be true for services that are public goods (services freely provide to all) such as ground water recharge, runoff mitigation, microclimate moderation, carbon sequestration, and so on. This represents a major challenge — finding ways to incorporate such substantial but hard-to-capture benefits into the calculus of private sector decision making.

2.7.3 Prospect

All of the approaches being used by business and industry illustrate how much the valuation of unique sociocultural amenities has been changing over time, and how these changes are becoming incorporated into decision processes. Both industry and government now need to take responsibility for managing long-term secondary (community-wide) impacts of development decisions, many of which stimulate changes in the relationship of the community to its environment. The scope of industry's responsibility in comparison to government's responsibility varies widely from one industry or one community to another, and from the regulatory regime of one country to another around the world. These realities impose a need for great sophistication in the measurements, analytic methods, and model projections discussed in the sections that follow.

2.8 DATA COLLECTION

As noted conceptually in section 2.7, there is a need for qualitative and quantitative measurement tools to help ascertain public values and attitudes about particular environmental decisions. A social impact assessment provides structure to data collection efforts in this arena. It is designed to help anticipate the effects on human communities of specified changes (McEvoy and Dietz 1977; Carley and Bustelo 1984:7; Interorganizational Committee on Guidelines and Principles 1994:12;

Burdge and Robertson 1998; Western and Lynch 2000:3). As such, it is a comparative exercise. It defines a baseline or current state, and anticipates the effects of various types of interventions (including "no action") to help enable decision makers to choose. It is very similar to an ecological risk assessment in concept, structure, and implementation.

Social impacts are measurable changes in a variety of dimensions of human communities, including many that are classed as "economic." Conduct of a social impact assessment requires that the assessor measure not just change in sociocultural indicators, but also such attributes of change as duration, intensity as functions of time and space, and the potential for reversibility. Selecting the indicators and relevant attributes to assess requires knowledge of sociocultural values, for it is through such values that importance would be assigned to the vector of the attribute by the community and the individuals that make it up. (Is a change in population worth worrying about? Is an increase or decrease good or bad? If a certain cultural practice needs to be eschewed for 5 years as a result of a particular decision, is that acceptable? If not, would 2 years be acceptable, and what would account for the difference?) Since the change in the value of an indicator can only be interpreted or understood in terms of a cultural "frame of interpretation" (Geertz 1973), an understanding of group-level values will help the decision maker appreciate the likelihood and magnitude of the response to his or her own actions, and help him or her determine the most appropriate path forward. Indeed, it has been argued that public participation in the identification and evaluation of impacts is central to a social impact assessment. It also should be noted that as participants' perceived stake in the outcomes of the decision becomes greater, values play a correspondingly greater role in community decision making than science (Renn et al. 1991).

There is a range of social indicators that could be used as part of a data collection template for a social impact assessment (Table 2.1). This list is not complete but illustrative. Each community, in conjunction with the specific anticipated decision or planned intervention, will develop a unique subset of these and other indicators for exploration and evaluation.

The reflexive process (Figure 2.2) engages the practitioner to determine which indicators to select. However, a problem arises when assessing the multiple social impacts of a proposed environmental change. The impacts may not be commensurable. Different members of the community may assign different importance or even valence to the same impacts, and impacts are likely to be distributed unevenly. The social impact assessment thus can inform a decision maker as to how (parts of) a community might respond to a decision, but it generally does not provide clear guidance to a "best" decision because it does not include methods for normalizing the impacts. One approach, derived from the theory of deliberative ethics described above, is to use community deliberation to examine trade-offs (Dietz 1987, 1988). Another is to use some generally accepted criteria, such as avoiding policies that constrain future choices or that polarize conflict (Freeman and Scott Frey 1986, 1990–1991). But the most commonly used approach is to apply a utilitarian logic to estimate changes in social value associated with alternative policies. This approach has by far the most theoretical and methodological development and is the subject of much of the rest of this volume.

TABLE 2.1
Social indicators in data collection for a social impact assessment

Population impacts
- Population change
- Influx or outflux of temporary workers
- Presence of seasonal (leisure) residents
- Relocation of individuals and families
- Dissimilarity in age, gender, racial, or ethnic composition

Community and infrastructure needs
- Change in community infrastructure
- Land acquisition and disposal
- Effects on known cultural, historical, sacred, and archeological resources

Community and institutional relationships
- Interest group activity
- Alteration in size and structure of local government
- Presence of planning and zoning activity
- Industrial diversification
- Enhanced economic inequities
- Change in employment of minority groups
- Change in occupational opportunities
- Formation of attitudes toward the project

Conflicts between residents and newcomers
- Presence of an outside agency
- Introduction of new social classes
- Presence of weekend residents
- Change in the commercial and/or industrial focus of the community

Political and social structures
- Changes in distribution of power and authority
- Changes in mechanisms for exercise of power and authority

Individual- and family-level impacts
- Disruption in daily living and movement patterns
- Alteration in family structure
- Disruption in social networks
- Change in leisure opportunities
- Dissimilarity in religious practices
- Perceptions of public health and safety

Source: Adopted from Interorganizational Committee on Guidelines and Principles (1994).

A social impact assessment can help inform the decision maker about both how a community will respond to a decision that will have ecological impact, and the value or importance the community places on these associated, anticipated sociocultural changes. Collecting data on values associated with ecological goods and services may be undertaken using one or a combination of qualitative and quantitative methods. The selection of appropriate data collection methods and approaches

Sociocultural Valuation of Ecological Resources

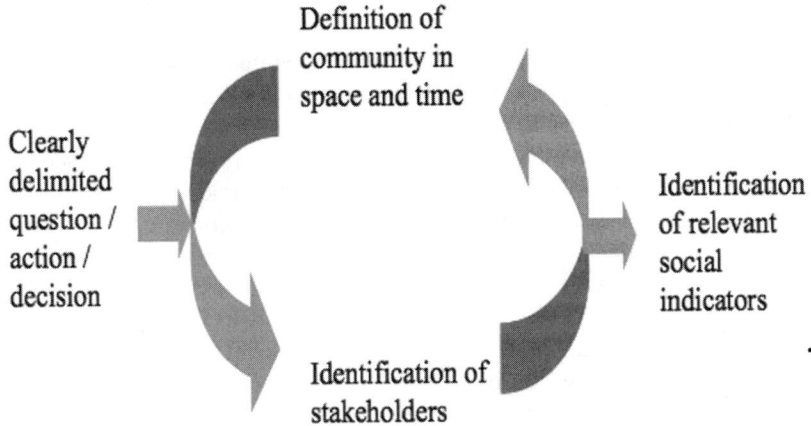

FIGURE 2.2 Selecting social indicators.

for ecologically based values depends, to a large extent, on the size and nature of the ecological goods and services issues under investigation. Likewise, the budget of the investigators determines the feasibility and complexity of valuation efforts. In any event, the social, cultural, and economic data needs as well as the valuation context dictate much about the relative attractiveness and utility of specific data collection and analysis methods.

At the start of any valuation effort, an understanding of the particular scope and circumstances of such an endeavor ought to precede investigators' answering threshold questions concerning the types of value and the beneficiaries to be considered. Such threshold questions, together with obvious constraints such as time and money, frame the appropriateness and reasonableness of alternative data collection methods. For example, a statewide mail survey for learning about public values associated with a small, isolated wetland ecosystem may seem inappropriate for a site-specific effort. The same survey effort may be more than appropriate, however, in the context of a statewide effort to protect endangered species dependent upon such small isolated wetlands. Likewise, the size and scope of stakeholder and other public participatory processes for collecting information on values depend, to a great extent, on the issue at hand and the political context. Consideration of a local wetland ordinance may warrant public meetings in its own locale, while consideration of a statewide wetland protection regulation may merit many larger public meeting across the state.

Financial constraints must also be taken into account. In any event, the collection of data on sociocultural values should be concerned about collecting data that are representative of the target populations. This may or may not mean that study results are generalizable outside the scope of a particular effort. However, all social, cultural, and economic valuation efforts benefit from transparency in their underlying assumptions regarding data collection and analyses. Being explicit about assumptions, constraints, and rationale for valuation approaches is important. That is, making clear what is and is not included in the social, cultural, and economic data collection and analysis processes goes a long way toward making those data most useful to decision makers and the public.

2.8.1 DATA COLLECTION PROCESSES

There has been a growing awareness over recent years that the manner of data elicitation can significantly impact and bias the type of data collected and hence the decisions that are made based on those data. Data collection process managers in multicultural communities or in communities that are heterogeneous on other dimensions (such as socioeconomic status) need to be sure that process structures, ranging from the time and place of meetings to conversational practices to the construction and administration of surveys, do not unduly favor and/or exclude any one segment of the affected population. The National Environmental Justice Advisory Council (1996) provides a checklist of value and process issues that can help a process manager ensure that participation is broad based and representative.

"Participatory processes" is a broad term that is often used to refer to data elicitation processes that involve a wide base of stakeholders or interested parties. The approach can involve a variety of data collection methods. Deliberative processes — a subset of participatory processes — focus on the collection of qualitative data through "deliberations" among interested parties. These often take the form of "public meetings" (often mandated by regulatory guidelines) or more elaborately constructed deliberative processes.

Effective participatory or stakeholder processes assume a pluralistic approach to decision making and a tradition of participatory process and its attendant structure. For example, participants in a workshop covering public participation in Poland noted that, under the Soviet regime, there was a history of direct relationship of the individual with the central government. The local government and civil sector, consisting of intermediate organizations and institutions such as NGOs, were nonexistent. This made Western participatory models very difficult to follow in many instances.

There are certain process issues that need to be considered regardless of what form the participatory process takes, several of which have been addressed earlier in this chapter. These include issues of definition of the relevant community, identification of relevant stakeholders, and appropriate construction of the process so that process concerns such as equity and access are addressed. Yosie and Herbst (1998) also note that issues such as measurement of process effectiveness and results and process quality management have yet to be addressed.

True deliberative processes also assume a transactional approach to policy making that is based on the democratic theories of Dewey and Habermas referred to earlier (Habermas 1970, 1993). That is, the desired end state is not acceptance or rejection of decisions made prior to engagement of stakeholders in the process, but the cooperative development of a solution that most completely satisfies the greatest number of participants. Most participative processes fall somewhat short of this ideal state.

Identifying the interested parties that are appropriate for inclusion is one of the primary areas where participative processes fail (Finney and Polk 1995; Glicken 1996, 1999). Generally the exclusion of appropriate parties from the dialogue is inadvertent; however, there are very few tools available to the process manager to help ensure that all affected parties are identified and the appropriate ones included.[9]

---

[9] See Glicken (1999) for discussion of one of these, a stakeholder "map" that builds on the relative importance of the identified groups vis-à-vis the issue in question, and their structural relationships.

The field of public participation — and, in particular, deliberative processes — is undergoing rapid change. For many years, practitioners used personal skills and ad hoc procedures to guide participation, while theorists considered the ethical benefits and requirements of participation. However, the last decade has seen the emergence of an empirical literature that carefully evaluates methods of public participation and tests hypotheses about process, context, and outcome (e.g., Tuler and Webler 1995, 1999; Chess and Purcell 1999; Beierle and Konisky 2000; Chess 2000; Webler 2000; Beierle 2002; Beierle and Cayford 2002). A US National Research Council (NRC) study is currently underway to evaluate the state of the science with regard to public participation and provide recommendations both for practice and for further research.[10]

2.8.2 QUALITATIVE AND QUANTITATIVE DATA

Social scientists categorize research methods or techniques for collecting social, cultural, and economic data into 2 broad categories — qualitative and quantitative methods. The former typically refers to such techniques as in-depth interviews, participant observation, and focus groups, while the latter most often refers to some sort of survey questionnaire or econometric analysis based on official records and data (e.g., hedonic analysis). The various approaches have their relative strengths and weakness. In both qualitative and quantitative analysis, great care must be taken in designing questions and interpreting the results. This is the subject of an immense and sophisticated literature in the social sciences. With proper care, results from qualitative and quantitative methods can provide deep and clear insights into the concerns and values of people who will be affected by a project or policy. Without such care, either type of method is unlikely to produce useful results. The sections following give a very brief overview of these methods; a little more detail can be found in Appendix 2.1.

The inherent differences in how data are reported and analyzed are an important distinction between qualitative and quantitative methods. Qualitative data typically take the form of transcripts, audio and video recordings, and researcher's notes. As such, these data require some sorting or coding process in order for investigators to distill the results and research findings. It is common practice for qualitative data analysis to make use of representative quotations to illustrate and support a particular "finding." In contrast, the data from survey questionnaires typically are numbers corresponding to answer choices to survey questions. These response variables are typically entered into a database and analyzed statistically. No one technique by itself is ideally suited to answer all of the research questions pertaining to the social, cultural, and economic values associated with ecosystems. It has become increasingly clear that iterative, multiple-method approaches that use both qualitative and quantitative methods seem most effective and promising for developing successful ecosystem valuation studies (Lupi et al. 2002; Kaplowitz et al. 2004).

A better understanding of some of the elements of alternative data collection methods may facilitate the identification of the most appropriate methods for a valuation task, as well as the inherent limitations associated with a particular undertaking. In addition, researchers in many fields (e.g., sociology) have long recognized the value of collecting social and cultural data using qualitative methods such as

---

[10] http://internationalacademies.org/gateway/international/1107.html. Accessed May 8, 2007.

cognitive interviews and focus groups (Morgan 1993; Krueger 1994; Weiss 1994; Morgan et al. 1998). Increasingly, survey researchers have been advised to consider using focus group interviews as well as individual interviews for pretesting and development of survey-based studies (e.g., Belson 1981; Bryman 1988; Mitchell and Carson 1989; Oppenheim 1992). Investigators, including economists and sociologists, assert the importance of qualitative analysis of people's perceptions of the problem in framing their investigation (e.g., Smith 1993; Foddy 1996). Qualitative methods also play an important role in the design and evaluation of effective survey instruments (Kaplowitz et al. 2004). Therefore, qualitative methods such as focus groups and individual interviews not only represent important methods as stand-alone tools but also more than bear consideration as design tools for improved surveys.

2.8.2.1 Methods for Eliciting Qualitative Data

Two principle qualitative research methods are focus groups and individual in-depth interviews. Expert panels can be treated as a special type of focus group. These methods share many of the same features — carefully designed sets of questions, trained moderator-interviewers, and an open-ended format for responses. However, the 2 methods differ, as their names suggest, in the size and number of respondents interacting with the moderator. Focus group interviews are conducted to allow researchers to learn about issues, concerns, and perceptions using a group discussion led by a moderator. It is generally accepted that focus group interviews rely on the dynamics of group interactions to reveal participants' similarities and differences of opinion as well as their understandings and beliefs (Krueger 1994; Morgan 1996, 1997; Knodel 1997). In contrast, individual qualitative interviews (also known as "in-depth and unstructured interviews") enable researchers to collect in-depth information from one respondent at a time and require the investigator to make comparisons with other interviews to determine similarities and differences (Oppenheim 1992; Weiss 1994; Knodel 1997). The relative strengths and weaknesses of particular qualitative research methods have "been more the subject of speculation than systematic research" (Morgan 1997). However, recent studies do show that these methods reveal different information that is complementary but not the same (Wight 1994; De Jong and Schellens 1998; Kaplowitz 2000; Kaplowitz and Hoehn 2001).

Resource economists and investigators in diverse fields have increasingly begun to use qualitative methods as comprehensive research tools and as components in designing and implementing reliable research efforts concerning natural resource values (e.g., Carson et al. 1994; Hoehn and Krieger 1994; Schwarz 1997; Lupi et al. 2002). As recommended by Carson (2000), sound nonmarket valuation studies require substantial survey development work. Carson (2000) recommends using focus groups and in-depth interviews to determine the plausibility and understandability of the environmental goods and services and the scenarios to be presented. Environmental economists do report some use of qualitative methods for the design, evaluation, and testing of questionnaires for valuing environmental resources (e.g., Mitchell and Carson 1989; Boyle et al. 1994; Carson et al. 1994; Chilton and Hutchinson 1999). Iterative use of focus groups and individual interviews in designing ecosystem valuation studies has been shown to be useful both as survey development tools and as free-standing valuation research tools (Kaplowitz et al. 2004).

2.8.2.2 Methods for Eliciting Quantitative Data

For the purpose of this chapter, surveys are quantitative methods for collecting social, cultural, and economic data on people's value associated with ecological resources. These surveys should, whenever possible and regardless of their type, conform to the highest standards and best practices for survey design and implementation. Suffice to say that the survey instruments themselves should be designed, evaluated, and pretested before implementation (see Presser et al. 2004) and they should be implemented so as to avoid sampling error and maximize response rates (see Dillman 2000). Beyond the foregoing caveats and references, it is beyond the scope of this volume to provide readers with a primer on survey development and implementation principles.

There are different types of surveys that relate to people's values of ecological resources. Social surveys or attitude surveys in the field of ecosystem studies attempt to collect information from the target populations concerning individuals' knowledge, understanding, and perception of ecosystem services and values. The goals of these types of surveys often include the assessment of stakeholder uses, understanding, and concerns regarding their ecosystems and ecosystem services, as well as an assessment of the relative acceptability of alternative management approaches. Ecological valuation, sometimes referred to as "nonmarket valuation," is a field of economics aimed at estimating economic values for changes in environmental and ecosystem services (Freeman 1993). Because environmental and natural resources are not typically traded on markets, economists have developed survey methods for environmental valuation based on individuals' preferences. Many of these methods are developed in greater detail in later chapters of this book.

2.8.2.3 Value of Multiple Methods

It seems best to employ a multiple-method approach for the design and evaluation of a study that elicits social, cultural, and economic values associated with ecosystems. While some environmental economics literature stresses the importance of survey development for choice studies (Adamowicz et al. 1998; Louviere et al. 2001) and for environmental valuation (Mitchell and Carson 1989; Carson 2000), few studies fully describe the processes used to design and evaluate valuation questionnaires. Instead, most published studies devote no more than a paragraph to the topic. Likewise, the major texts in the fields for choice models (Louviere et al. 2001; and see Mitchell and Carson 1989 for environmental valuation) state the importance of survey development. Recent work demonstrates the value and importance of using multiple methods for survey design and environmental valuation (Tourangeau et al. 2000; Kaplowitz and Hoehn 2001; Kaplowitz et al. 2004; Presser et al. 2004). Practitioners in the environmental valuation field stand to benefit from the research and methodological advances on the use of multiple methods.

2.9 CONCLUSION

Developing an understanding of what it means to value an ecological resource is a complex process. It is highly context dependent, depending as much on the type of decision in play as on the type of resource in question. The value will vary from person to person, from place to place, and over time.

The decision maker needs to be very clear as to the type of question for which the value has import. Decisions regarding the location of a new industrial facility have different ramifications than do decisions regarding the licensing of a new pesticide. The range of interested parties is different, the geographic scope of impact of the targeted activity is different, and the interests at play may be significantly different. These differences will lead to the identification of different communities of interest. Budget and time constraints on data collection also will come into play. All these factors will impact the identification of significant values, which, in turn, may influence the data collection methods employed and the type of data collected.

REFERENCES

Adamowicz, W, Boxall, P, Louviere, L, Swait, Williams, M. 1998. Stated preference methods for valuing environmental amenities. In: Batemen I, Willis, K, editors. Valuing environmental preferences: theory and practice of the contingent valuation method in the US, EC and developing countries. Oxford (UK): Oxford University Press.

Arrow, KJ, Cropper, ML, Eads, GC, Hahn, W, Lave, LB, Noll, RG, Portney, PR, Russell, MR, Schmalensee, R, Smith, VK, Stavins, RN. 1996. Is there a role for benefit-cost analysis in environmental, health, and safety regulation? Science 272(5259):221–222.

Arrow, K., Solow, Leamer, RE Portney, P, Rader, R, Schuman, H. 1993. Report of the NOAA panel on contingent valuation. Federal Register 58(10):4601–4614.

Axelrod R. 1984. The evolution of cooperation. New York (NY): Basic Books. 256 p.

Babbie, EJ. 1990. Survey research methods. Belmont (CA): Wadsworth.

Beierle TC. 2002. Democracy on-line: an evaluation of the national public dialogue on public involvement in EPA decisions. Washington (DC): Resources for the Future. 80 p.

Beierle TC, Cayford J. 2002. Democracy in practice: public participation in environmental decisions. Washington (DC): Resources for the Future. 148 p.

Beierle TC, Konisky DM. 2000. Values, conflict and trust in participatory environmental planning. J Policy Anal Manag 19:587–602.

Belson, W A. 1981. The design and understanding of survey questions. London (UK): Gower.

Boyle, K, Desvousges, JWH, Johnson, FR, Dunford, R, Hudson, SP. 1994. An investigation of part-whole biases in contingent valuation studies. Journal of Environmental Economics and Management 27:64–83.

Boxall, PC, Adamowicz VL. 1999. Understanding heterogeneous preferences in random utility models: the use of latent class analysis. Edmonton (AB): Department of Rural Economy, University of Alberta. 47 p.

Brown L, editor. 1993. The new shorter Oxford English dictionary on historical principles. Oxford (UK): Clarendon Press. 3801 p.

Bryman, Alan. 1988. Quantity and quality in social research. Bulmer, M, editor. Vol. 18, Contemporary Social Research. London (UK): Unwin Hyman.

Burdge RJ, Robertson RA. 1998. Social impact assessment and the public involvement process. In: Burdge RJ, editor, A conceptual approach to social impact assessment: collection of writings. Rev. ed. Middleton (WI): Social Ecology Press, p 183–192.

Carley MJ, Bustelo ES. 1984. Social impact assessment and monitoring: a guide to the literature. Boulder (CO): Westview Press. 250 p.

Carson R. 1962. Silent spring. Boston (MA): Houghton-Mifflin. 368 p.

Carson, RT, Mitchell, RC. 1993. The issue of scope in contingent valuation. American Journal of Agricultural Economics 75:1263–1267.

Carson, RT, Hanemann, WM, Kopp, RJ, Krosnick, A, Mitchell, RC, Presser, S, Ruud, PA, Smith, VK. 1994. Prospective interim lost use value due to DDT and PCB contamination in the southern California bight. La Jolla (CA): Natural Resource Damage Assessment, Inc.

Carson, RT. 2000. Contingent valuation: a user's guide. Environmental Science & Technology 34(8):1413–1418.

Carson, RT., Hanemann, WM, Kopp, RJ, Krosnick, JA, Mitchell, RC, Presser, S. 1998. Referendum design and contingent valuation: The NOAA panel's no-vote recommendation. Review of Economics and Statistics 80(3):484–487.

Chess C. 2000. Evaluating environmental public participation: methodological questions. J Environ Plan & Mgmt 43:769–784.

Chess C, Purcell K. 1999. Public participation and the environment: do we know what works? Environ Sci Technol 33:2685–2692.

Chilton, S M, Hutchinson, WG. 1999. Exploring divergence between respondent and researcher definitions of the goods in contingent valuation studies. Journal of Agricultural Economics 50(1):1–16.

Cramer JC, Dietz T, Johnston R. 1980. Social impact assessment of regional plans: a review of methods and a recommended process. Policy Sci 12:61–82.

De Jong, Menno, Schellens, PJ. 1998. Focus groups or individual interview? A comparison of text evaluation approaches. Technical Communication 45(1):77–88.

Delbecq AL, Van de Ven AH, Gustafson DH. 1975. Group techniques for program planning: a guide to the nominal group and Delphi Process. Glenview (IL): Scott-Foresman. 174 p.

Devall B. 1986. Deep ecology. Layton (UT): Gibbs Smith Publishers. 267 p.

Dewey J. 1923. The public and its problems. New York (NY): Henry Holt. 236 p.

Dietz T. 1987. Theory and method in social impact assessment. Sociol Inq 57:54–69.

Dietz T. 1988. Social impact assessment as applied human ecology: integrating theory and method. In: Borden R, Jacobs J, Young GR, editors, Human ecology: research and applications. College Park (MD): Society for Human Ecology, p 220–227.

Dietz T. 1994. What should we do? Human ecology and collective decision making. Hum Ecol Rev 1:301–309.

Dietz T, Kalof L, Stern PC. 2002. Gender, values and environmentalism. Soc Sci Quart 83:353–365.

Dietz T, Ostrom E, Stern PC. 2003. The struggle to govern the commons. Science 302:1907–1912.

Dietz T, Pfund A. 1988. An impact identification method for development program evaluation. Policy Stud Rev 8:137–145.

Dietz T, Stern PC. 1998. Science, values and biodiversity. BioScience 48:441–444.

Dietz T, Stern PC, editors. 2002b. New tools for environmental protection: education, information and voluntary measures. Washington (DC): National Academy Press. 368 p.

Dillman, DA. 2000. Mail and Internet Surveys: The tailored design method, 2nd ed. New York (NY): John Wiley & Sons, Inc.

Ditz D, Ranganathan J, Banks RD, editors. 1995. Green ledgers: case studies in corporate environmental accounting. Baltimore (MD): World Resources Institute. 181 p.

Dryzek JS. 1994. Ecology and discursive democracy: beyond liberal capitalism and the administrative state. In: O'Conner M, editor, Is capitalism sustainable? Political economy and the politics of ecology. New York (NY): Guilford Press, p 176–197.

Dunlap RE, Grieneeks JK, Rokeach M. 1983. Human values and pro-environmental behavior. In: Conn DW, editor, Energy and mineral resources: attitudes, values and public policy. Washington (DC): American Association for the Advancement of Science, p 145–168.

Durkheim E. 1964. The rules of sociological method. 8th ed. Solovay SA, Mueller JH, translators. Catlin GEG, editor. New York (NY): Free Press. 272 p.
Erekson H, Krehbiel TC, Ohl A. 1999. Sustainability and business management systems. In: Loucks OL, Erekson H, Bol JF, Gorman RF, Johnson PC, Krehbiel TC, editors, Sustainability perspectives for resources and business. Boca Raton (FL): Lewis Publishers, p 139–166.
Erekson H, Leavitt A. 1999. From earth to the table: Fetzer Vineyards and its Bonterra Wines. Case Study Number 4. In: Loucks OL, Erekson H, Bol JF, Gorman RF, Johnson PC, Krehbiel TC, editors, Sustainability perspectives for resources and business. Boca Raton (FL): Lewis Publishers, p 249–269.
Ernst H. 2003. Chesapeake Bay blues science, politics and the struggle to save the bay. Lanham (MD): Rowman & Littlefield. 224 p.
Finney C, Polk RE. 1995. Developing stakeholder understanding, technical capability, and responsibility. New Bedford Harbor Superfund Forum. Environ Impact Assess 15:517–541.
Fiorino DJ. 1990. Citizen participation and environmental risk: a survey of institutional mechanisms. Sci Technol Hum Val 15:226–243.
Foddy, W. 1996. The in-depth testing of survey questions: a critical appraisal of methods. Quality & Quantity 30:361–370.
Fowler, F J, Jr. 1995. Improving survey questions: design and evaluation. Vol. 38, Applied Social Research Methods Series. Thousand Oaks (CA): Sage Publications.
Freeman AM III. 1993. The measurement of environmental and resource values. Washington (DC): Resources for the Future. 496 p.
Freeman DM, Scott Frey R. 1986. A method for assessing the social impacts of natural resource policies. J Environ Mgmt 23:229–245.
Freeman DM, Scott Frey R. 1990–1991. A modest proposal for assessing social impacts of natural resource policies. J Environ Syst 20:375–404.
Garfinkel H. 2002. Ethnomethodology's program: working out Durkheim's aphorism. Lanham (MD): Rowman & Littlefield. 320 p.
Geertz C. 1973. The interpretation of cultures: selected essays by Clifford Geertz. New York (NY): Basic Books. 460 p.
Gibson CC, Ostrom E, Ahn TK. 2000. The concept of scale and the human dimensions of global change: a survey. Ecol Econ 32:217–239.
Glicken J. 1996. Stakeholder mapping and environmental risk assessment: developing socially appropriate solutions to environmental problems. Presentation at American Anthropological Association 95th annual meeting, San Francisco, CA.
Glicken J. 1999. Effective public involvement in public decisions. Sci Comm 20(3):298–327.
Glicken J. 2000. Getting stakeholder participation "right": A discussion of participatory process and possible pitfalls. Environ Sci & Pol 3:305–310.
Glicken J, Fairbrother A. 1998. Environment and social values. Hum Ecol Risk Assess 4:779–786.
Goffman E. 1959. The presentation of self in everyday life. New York (NY): Anchor Books Doubleday. 259 p.
Goffman E. 1974. An essay on the organization of experience: frame analysis. Boston: Northeastern University Press. 600 p.
Gorman RF. 2000. Valuation and reporting. In: Loucks OL, Erekson H, Bol JF, Gorman RF, Johnson PC, Krehbiel TC, editors, Sustainability perspectives for resources and business. Boca Raton (FL): Lewis Publishers, p 105–138.
Habermas J. 1970. Towards a rational society. Boston (MA): Beacon Press. 132 p.
Habermas J. 1993. Justification and application: remarks on discourse ethics. Cambridge (MA): MIT Press. 229 p.

Hobbes T. (1651) 1998. Leviathan. New York (NY): Oxford University Press. 508 p.

Hoehn, JP, Krieger, D, Kaplowitz, MD. 1999. Estimating benefits of water and wastewater investments: residential demand in Cairo. International Review of Comparative Public Policy 11, 156-176.

Interorganizational Committee on Guidelines and Principles for Social Impact Assessment. 1994. Guidelines and principles for social impact assessment. NMFS-F/SPO-16. Washington (DC): US Department of Commerce, NOAA Technical Memo. http://www.nmfs.noaa.gov/sfa/social_impact_guide.htm. Accessed July 6, 2002.

Jaeger C, Renn O, Rosa EA, Webler T. 2001. Risk, uncertainty and rational action. London (UK): Earthscan. 324 p.

Kalof L, Dietz T, Guagnano GA, Stern PC. 2002. Race, gender and environmentalism: the atypical values and beliefs of white men. Race, Gender & Class 9:1–19.

Kaplowitz MD. 2000. Statistical analysis of sensitive topics in group and individual interviews. Qual & Quant: Intl J Method 34(4):1–12.

Kaplowitz MD, Hoehn JP. 2001. Do focus groups and personal interviews reveal the same information for natural resource valuation? Ecol Econ 36:237–247.

Kaplowitz, MD, Lupi F, Hoehn JP. 2004. Multiple-methods for developing and evaluating a stated choice survey to value wetlands. In: Presser S, Rothgeb JM, Couper MP, Lessler JT, Martin E, Martin J, Singer E, editors, Methods for testing and evaluating survey questionnaire. Hoboken (NJ): John Wiley. p 503–524.

Kluckholn C. 1952. Values and value-orientation in the theory of action: an exploration in definition and classification. In: Parsons T, Shils E, editors, Toward a general theory of action. Cambridge (MA): Harvard University Press, p 388–433.

Knodel, J. 1997. A case for nonanthropological qualitative methods for demographers. Population and Development Review 23(4):847–853.

Krueger RA. 1994. Focus groups: a practical guide for applied research. 2nd ed. Thousand Oaks (CA): Sage. 272 p.

Leopold A. 1949. A sand county almanac and sketches here and there. New York (NY): Oxford University Press.

Levitt AR. 1994. The interaction of human and ecological systems: considerations in the establishment of ecologically sensitive reserves. MS thesis, Miami University. 48 p.

Loucks DL, Erekson H, Bol JW, Gorman RF, Jackson, PC, Krehbiel TC, editors. 1999. Sustainability perspectives for resources and business. Boca Raton (FL): Lewis Publishers. 373 p.

Louviere, JJ, Hensher, DA, Swait, JD. 2001. Stated choice methods: analysis and applications. Cambridge (UK): Cambridge University Press.

Lovelock J. 1979. Gaia: a new look at life on earth. Oxford: Oxford University Press. 176 p.

Maxwell JA. 1996. Qualitative research design: an interactive approach. Applied Social Research Methods Series Vol. 41. Thousand Oaks (CA): Sage. 192 p.

Mayr E. 1959. Typological versus population thinking. In: Meggers BJ, editor, Evolution and anthropology: a centennial appraisal. Washington (DC): Anthropological Society of Washington. p 409–412.

McEvoy J III, Dietz T, editors. 1977. Handbook of environmental planning: the social consequences of environmental change. New York (NY): Wiley Intersciences. 323 p.

McLaughlin P. 2001. Towards an ecology of social action: merging the ecological and constructivist traditions. Hum Ecol Rev 8:12–28.

Miller AS. 1991. Gaia connections: an introduction to ecology, ecoethics, and economics. Savage (MD): Rowman & Littlefield. 324 p.

Mitchell, RC, Carson, RT. 1989. Using surveys to value public goods: the contingent valuation method. Washington (DC): Resources for the Future.

Morgan DL. 1997. Focus groups as qualitative research. Vol. 16. Thousand Oaks (CA): Sage. 92 p.
Morgan, DL. 1993. Successful focus groups: advancing the state of the art. Newbury Park (CA): Sage Publications.
Morgan, DL. 1996. Focus groups. In: Hagan, J, Cook, KS, editors. Annual Review of Sociology, Palo Alto (CA): Annual Reviews.
Morgan, DL, Krueger, RA, Scannell, AU, King, JA. 1998. Focus group kit. Thousand Oaks (CA): SAGE Publications.
Naess A. 1972. The shallow and the deep, long-range ecology movement. Inquiry 16:95–100.
Nash R. 1967. Wilderness and the American mind. Rev. edition. New Haven (CT): Yale University Press. 426 p.
Oppenheim, A N. 1992. Questionnaire design, interviewing and attitude measurement, new ed. New York (NY): Pinter Publishers.
National Environmental Justice Advisory Committee. 1996. The model plan for public participation. EPA-300-K-00-001. Washington (DC): USEPA. 20 p.
Parsons T. 1949. The structure of social action: a study in social theory with special reference to a group of recent European writers. Glencoe (IL): The Free Press. 776 p.
Pearce DW, Turner RK. 1990. Economics of natural resources and the environment. Baltimore (MD): Johns Hopkins University Press.
Powney, J, Watts, M. 1987. Interviewing in educational research. London (UK): Routledge and Kegan Paul.
Presser S, Rothgeb JM, Couper MP, Lessler JT, Martin E, Martin J, Singer E. 2004. Methods for testing and evaluating survey questionnaire. Hoboken (NJ): John Wiley.
Rea LM, Panker RA. 1992. Designing and conductions survey research. San Francisco (CA): Jossey-Bass.
Renn O, Webler T, Johnson BB. 1991. Public participation in hazard management: the use of citizen panels in the US Risk — Iss in Health & Safety 197:197–226.
Renn O, Webler T, Rakel H, Johnson B, Dienel P. 1993. A three-step procedure for public participation in decision making. Policy Sci 26:189–214.
Renn O, Webler T, Wiedemann P, editors. 1995. Fairness and competence in citizen participation: evaluating models for environmental discourse. Dordrecht (the Netherlands): Kluwer Academic. 408 p.
Richerson PJ. 1977. Ecology and human ecology: a comparison of theories in the biological and social sciences. Am Ethnol 4:1–26.
Robson, C. 1993. Real world research. Oxford (UK): Blackwell.
Rokeach M. 1968. Beliefs, attitudes and values: a theory of organization and change. San Francisco (CA): Jossey-Bass. 289 p.
Rokeach M. 1973. The nature of human values. New York (NY): Free Press. 447 p.
Rosa E, McCright AM, Renn O. 2000. The risk society: theoretical frames and state management challenges. Paper presented at annual conference of the Society of Risk Analysis — Europe, May, Edinburgh, Scotland.
Rose C. 2002. Common property, regulatory property and environmental protection: comparing community-based management to tradable environmental allowances. In: Ostrom E, Dietz T, Dolsak N, Stern PC, Stonich S, Weber E, editors, The drama of the commons. Washington (DC): National Academy Press. p 233–258.
Sagoff M. 1988. The economy of the earth. Cambridge (UK): Cambridge University Press. 271 p.
Sagoff M. 2000. Environmental economics and the conflation of value and benefit. Environ Sci Technol 34:1426–1432.
Schwarz, N. 1997. Cognition, communication, and survey measurement. In: Kopp, RJ, Pommerehne, W, Schwarz, N, editors. Determining the value of non-marketed goods. Boston (MA): Kluwer-Nijhoff.
Schwartz SH. 1987. Toward a universal psychological structure of human values. J Pers Soc Psychol 53:550–562.

Schwartz SH. 1992. Universals in the content and structure of values: theoretical advances and empirical tests in 20 countries. Adv Exp Soc Psychol 25:1–65.

Schwartz SH, Bilsky W. 1990. Toward a theory of the universal content and structure of values: extensions and cross-cultural replications. J Pers Soc Psychol 58:878–891.

Schuman H, Presser DS. 1996. Questions and answers in attitude surveys: experiments on questions form, wording, and content. Thousand Oaks (CA): Sage Publications.

Smart B. 1992. Beyond compliance: a new industry view of the environment. Washington (DC): World Resources Institute. 285 p.

Smith, VK. 1993. Nonmarket valuation of environmental resources: an interpretive appraisal. Land Economics 69:1–26.

Sober E. 1980. Evolution, population thinking, and essentialism. Philos Sci 47:350–383.

Stern PC. 2000. Toward a coherent theory of environmentally significant behavior. J Soc Issues 56:407–242.

Stern PC, Dietz T. 1994. The value basis of environmental concern. J Soc Issues 50:65–84.

Stern PC, Dietz T, Abel T, Guagnano GA, Kalof L. 1999. A social psychological theory of support for social movements: the case of environmentalism. Hum Ecol Rev 6:81–97.

Stern PC, Dietz T, Kalof L. 1993. Value orientations, gender and environmental concern. Environ Behav 25:322–348.

Stern PC, Fineberg H, editors. 1996. Understanding risk: informing decisions in a democratic society. Washington (DC): National Academy Press. 250 p.

Tietenberg T. 2002. The tradable permits approach to protecting the commons: what have we learned? In: Ostrom E, Dietz T, Dolsak N, Stern PC, Stonich S, Weber E, editors, The drama of the commons. Washington (DC): National Academy Press. p. 197–232.

Tourangeau, R, Rips, LJ, Rasinski, K. 2000. The psychology of survey response. Cambridge (UK): Cambridge University Press.

Tuler S, Webler T. 1995. Process evaluation for discursive decision making in environmental and risk policy. Hum Ecol Rev 2:62–71.

Tuler S, Webler T. 1999. Voices from the forest: what participants expect of a public participation process. Soc Natur Resour 12:437–453.

Udehn L. 2002. The changing face of methodological individualism. Ann Rev Soc 28:479–507.

US National Research Council. 1999. Perspectives on biodiversity: valuing its role in an ever-changing world. Washington (DC): National Academy Press. 153 p.

Ward, KM., Duffield, JW. 1992. Natural resource damages: law and economics, environmental law library. New York (NY): Wiley Law Publications.

Webler T. 1993. Habermas put into practice: a democratic discourse for environmental problem solving. In: Wright SD, Dietz T, Borden R, Young G, Guagnano G, editors, Human ecology: crossing boundaries. Ft. Collins (CO): Society for Human Ecology. p 60–72.

Webler T. 2000. Fairness and competence in citizen participation: theoretical reflections from a case study. Admin Soc 32:566–595.

Weiss, RS. 1994. Learning from strangers: the art and method of qualitative interview studies. New York (NY): The Free Press.

Western J, Lynch M. 2000. Overview of the SIA process. In: Goldman LR, editor, Social impact analysis: an applied anthropology manual. New York (NY): Berg. p 35–62.

Whitfield S, Dietz T, Rosa EA. 1999. Environmental values, risk perception and support for nuclear technology. Fairfax (VA): Human Ecology Research Group, George Mason University.

Wight, D. 1994. Boys' thoughts and talk about sex in a working class locality of Glasgow. Sociological Review. 42:702–737.

Wilson EO. 1975. Sociobiology: the new synthesis. Cambridge (MA): Harvard University Press. 697 p.

Wilson EO. 2000. Consilience: the unity of knowledge. New York (NY): Alfred A. Knopf. 384 p.

Yohe G, Toth FL. 2000. Adaptation and the guardrail approach to tolerable climate change. Climat Chg 45:103–128.

Yosie TF, Herbst TD. 1998. Using stakeholder processes in environmental decisionmaking: an evaluation of lessons learned, key issues, and future challenges. Washington (DC): Ruder Finn.

APPENDIX 2.1

A.1 DATA COLLECTION PROCESSES

The respondent identification process begins with appropriate identification of the communities in question. This takes us back to the earlier discussion of whose values are on the table. Deep ecologists and others would say that as all ecological systems are interrelated, any decision with an ecological impact impacts us all. Therefore, we all are interested parties in any decision. Obviously, this is a little unwieldy in operation, so some boundaries need to be drawn. Again, we refer to the Arabian Gulf and Madagascar case studies at the end of this volume (Case Studies 4 and 3, respectively) for examples of how some of these "nested" value sets were operationalized in actual ecological valuation scenarios.

Just as the community is defined relative to a particular problem and proposed action, so are stakeholders or interested parties. Interested parties — just as "communities" — are defined relative to an issue. They can range from corporate entities (that is, organizations that have an identity separate from the individuals who comprise them and so can imbue designated representatives to speak for the whole) through membership groups (much more loosely constituted groups such as neighborhood associations) to individuals. In all cases, they are those who perceive their interests to be impacted in some way by the action in question. Again, stakeholder maps, drawing on the principles used to develop folk taxonomies, can be useful tools to ensure that relevant relationships and organizations are recognized.

A.1.1 Methods for Collecting Qualitative Data

A.1.1.1 Focus Groups

Focus groups may help investigators learn what it is about ecosystems, ecosystem services, and ecosystem characteristics that matter to people. They may be used to help scope out the territory and concepts for a valuation effort — that is, frame subsequent research — or may be conducted as independent research efforts. The size of focus group sessions is conditioned by the desire for them to be "small enough for everyone to have opportunity to share insights and yet large enough to provide diversity of perceptions" (Krueger 1994:17). Typically, focus group sessions have about 6 or 7 participants plus a professional moderator. Participants recruited to participate in a focus group are usually asked to meet at a centrally located facility where they will be greeted by members of the research team and offered some refreshments before the group session begins. Ideally, focus groups will be held with randomly recruited members of the target population. A professional moderator using a discussion guide prepared by the investigators in conjunction with the moderator should lead focus group discussions. The discussion guide should frame the discussion so that the issues, concepts, and behavior of importance to the study may be considered by the focus group participants. Typically, focus group sessions are video- and audio-recorded to allow for subsequent in-depth analysis of discussion content (Krueger 1994; Morgan 1996, 1997). At the conclusion of the session, participants generally receive a small customary honorarium ($40) to thank them for their participation in the focus group study.

Following the maxim that one needs as many sessions so as nothing new is being learned in the last session (Maxwell 1996; Morgan 1997), researchers should budget to conduct enough focus groups until no major new information is revealed. Focus group sessions should follow a detailed discussion guide to lead respondents through several topics. For ecological valuation studies, such topics likely include ecological resources of importance to people, prior knowledge of the ecological resources, cognizance of current policies, and reaction to possible change scenarios. Doing so can help the investigators learn how the target population (e.g., general public) thinks about ecological resources, identify information gaps, and frame subsequent valuation tasks (e.g., designing information treatments for survey questionnaires).

There have also been suggestions that the selection of individuals to participate in focus groups and the process used to conduct the groups should be attentive to what is known about the social psychology of group dynamics and the issues to be addressed by the group (Cramer et al. 1980; Dietz 1988). Delbecq et al. (1975) have developed a "nominal group technique" that has been adopted to social impact assessment (Dietz and Pfund 1988). Renn et al. (1993) have also suggested a group process for eliciting community views around risk. These methods attempt to deploy the structure typical of quantitative methods while allowing for the open-endedness of qualitative methods.

A.1.1.1.1 Expert Panels

In addition to learning from the general public about their social, cultural, and economic values for ecosystems and ecosystem services, investigators can benefit from learning from scientific experts about their perceptions and understanding of such values. Of course, the various methods discussed in this chapter and volume may be tailored for scientific participants. However, one especially useful technique for learning from scientific experts closely resembles focus groups. The use of expert panels — at the national scale such as NRC panels, at the professional level such as SETAC Pellston panels, or at the local level such as state science advisory panels — provides a forum in which key personnel can discuss ecological, social, and economic concepts relating to a project, problem, or program focus. At the regional scale, these sorts of groups of experts can include state program coordinators, leading scientists from appropriate agencies, experts from academia, as well as scientists from NGOs. Like focus groups, these experts likely need a moderator, but the agenda can be transparent with the agreed-upon tasks to include 1) helping the research team; 2) clarifying pertinent social, cultural, and economic values associated with the ecosystem and ecosystem service under investigation; 3) understanding appropriate local, statewide, and national protection and mitigation policies; and 4) identifying valuation needs beyond the reach of conventional regulatory means. Use of such a panel of experts may help to provide scientific expertise and guidance to the researchers. Furthermore, vetting of research design choices including design elements with these experts can help to ensure the soundness of subsequent data collection methods and analyses.

Of course, there are several kinds of expertise, of which scientific and technical expertise is only one (Dietz 1987). Members of local communities and those who are politically engaged in issues also bear expertise. Indeed, recent approaches to risk assessment, valuation, and environmental decision making place a strong emphasis on methods and procedures that facilitate dialogue among community members and scientific and technical experts (Stern and Fineberg 1996; Dietz and Stern 1998). The challenge in these approaches is to find ways to respect and integrate multiple understandings of a problem into analysis and valuation.

A.1.1.2 Individual Interviews

"Individual interviews" may refer to a range of methods of collecting information from people. While some might include a face-to-face survey in this group of methods, for our purposes "individual interviews" refer to systematic one-on-one interviews that are open-ended rather than based on a structured questionnaire. Individual interviews may be semistructured interviews (Powney and Watts 1987; Robson 1993), or they may be much more exploratory in nature (Weiss 1994). Individual interviews — also called qualitative interviews, cognitive interviews, intensive individual interviews, and debriefing interviews (Fowler 1995; Tourangeau et al. 2000) — resemble focus groups save for the absence of group dynamics. In fact, it is precisely the absence of group dynamics that makes individual qualitative interviews so useful.

Like focus groups, individual interviews may be used for the scoping of issues. However, compared to the costs of doing focus groups, individual interviews are a relatively expensive means for collecting such information from members of the general public. For example, a scoping focus group of 8 participants requires only a single 1–2-hour session with 1 moderator. One would need 8 separately scheduled individual interviews with a moderator to collect scoping information from the same 8 members of the public. Research does suggest that individual interviews are particularly well suited for use in evaluating the details of individuals' thinking about ecological issues, values, and other specific tasks. In particular, cognitive interviews have been recognized as extremely useful in evaluating the design and efficacy of survey questionnaires (e.g., Kaplowitz et al. 2004)

One use of individual interviews is to develop basic insights that can hone and focus survey instruments as well as provide useful information per se. Participants may be given time to independently complete a version or draft of the valuation questionnaire. Upon completion of the questionnaire, trained interviewers may lead respondents through a scripted cognitive interview to evaluate their comprehension of the different aspects of the instrument. The interview script typically begins with the more open-ended approach to ensure that respondents' general thoughts and concerns were not influenced by the more directed questions. These interviews, which generally take 30 to 45 minutes, focus on respondents' comprehension and thought process as they engaged in the valuation questions. Similarly, respondents may be queried about their understanding of the information presented in the survey in support of the attributes and context of the valuation exercise. The interview script may be structured so that the interviews begin in a semistructured way with very open-ended questions to elicit nondirected feedback on the questionnaires in a casual, talkative setting. The script may also progress through increasingly directed sets of questions. For example, respondents may be asked to talk about how they made their choices. Following this discussion, respondents may be directed to talk about the effect that each of the specific attributes in the table had on their choice. The interviews typically end with some very specific questions about aspects of the interview.

A.1.2 METHODS FOR COLLECTING QUANTITATIVE DATA

A.1.2.1 Attitude Surveys

These surveys may be conducted face-to-face with an interviewer, or they may be conducted over the telephone, via the mail, or via Web-based survey questionnaires. While face-to-face interviews may be used, typically researchers use a self-administered mail survey questionnaire There has been a great deal written on designing and conducting survey research (e.g., Babbie 1990; Oppenheim 1992; Rea and Parker 1992; Schuman and Presser 1996; Dillman 2000; Tourangeau et al. 2000). It is beyond the scope of this chapter to provide detailed descriptions of survey research methods. It suffices to say that all good attitude surveys address such issues as identification of target population(s), appropriate sample selection, instrument design and pretesting, survey implementation, and data analysis. In addition to demographic data collected from each respondent, attitude surveys typically gather respondents' ratings of their degree of agreement or disagreement with a variety of statements. Use of such Likert-type scale measures (e.g., 5-point ratings) of respondents' attitudes is fairly common in many social science studies. Survey instruments may also be used to collect use information (e.g., frequency and duration of visits, and substitutes) as well as characteristics of respondents (users). Well-designed and implemented surveys have the potential for being able to generate quantitative and generalizable results.

A.1.2.2 Valuation Surveys

A panel chaired by Nobel laureate scholars affirmed the validity and usefulness of survey questionnaires for environmental valuation (Arrow et al. 1993). Arrow et al. (1993) and others stress the importance of ensuring that 1) respondents understand the environmental goods and services to be valued; 2) respondents receive reliable, uniform, and complete sets of information about the environmental goods and services to be valued; and 3) respondents are presented with a realistic and acceptable choice context. The methodological challenge of designing environmental valuation questionnaires has been recognized for some time (Mitchell and Carson 1989; Carson and Mitchell 1993; Carson et al. 1998). "Producing a good [environmental valuation] survey instrument requires substantial development work" (Carson 2000:1415). Information on the economic benefits of environmental quality may be used in cost–benefit analyses of environmental policies (Arrow et al. 1996) and legal cases involving natural resource damages (Ward and Duffield 1992).

As will be explained elsewhere in this volume, nonmarket valuation questionnaires typically present respondents with information about the attributes (e.g., size, type, quality, and cost) of particular environmental goods or services. The questionnaires also inform respondents about the choice context and implications of possible trade-offs. They then ask respondents to choose between alternative bundles of goods and services. In using such questionnaires, all respondents receive identical information about the attributes, policies, and choice context. That is, respondents receive the same information treatments but make choices among alternative scenarios with differing attribute levels. Respondents are asked to select their preferred "outcome." Statistical analysis of the attribute trade-offs implicit in respondents' choices reveals underlying economic values associated with the goods and services. The analysis of individuals' informed trade-offs reveals the public's value for the environmental and natural resource services in question (see, e.g., Lupi et al. 2002).

3 Integrating Economics and Ecological Assessment

*Richard C. Bishop, Joshua Lipton,
Michael Margolis, Norman Meade,
George L. Peterson, and Alan Randall*

CONTENTS

3.1 Introduction 45
3.2 The Many Roles of Economics in Environmental Policy Analysis 46
3.3 Economic Valuation for Environmental Decision Making 48
3.4 Theory-Based Critiques of Valuation from Within Economics 50
3.5 Practical Problems in Measuring Benefits and Costs 52
3.6 Critiques from Outside Economics: Enduring Value, Intrinsic Value,
 and Deep Ecology 54
3.7 Summary and Conclusions 56
References 56

3.1 INTRODUCTION

Economics is the science of the allocation of resources, which are defined broadly to encompass labor and capital as well as environmental and other natural resources. When there are not enough resources to go around, choices must be made, and economics is the science of choosing.

Some elaboration on our use of the term "science" in defining economics may be helpful given the interdisciplinary audience of this book. We believe that economics falls within the *American Heritage Dictionary*'s definition of science: "The observation, identification, description, experimental investigation, and theoretical explanation of phenomena." At the same time, just as there are large theoretical and methodological differences between the different physical and biological sciences, so too does economics take on its own distinctive character. Natural sciences such as ecology focus mostly on description, explanation, and prediction of natural phenomena. Economics has a counterpart known as "positive" economics. A familiar example is the theory of supply and demand. If, for example, the demand for a good increases and all else remains constant, economics predicts that the price of that good will rise and seeks to explain why. More so than the natural sciences, though, economics has a "normative" side. At the center of applied normative economics is benefit–cost analysis (BCA). For example, BCA might be applied to a proposal to strengthen air pollution regulations. Here economics would be normative in the sense that it is not attempting to predict whether or not air regulations will be strengthened, but whether, from an economic point of view, they ought to be strengthened. We will touch on both normative and positive economics in this chapter, but the emphasis will be on the normative side.

Two principal questions addressed by normative economics are economic efficiency and economic equity. In the arena of public policy, economic efficiency analysis asks whether a particular choice will increase the net aggregate wealth, or well-being, of society, without regard to who gains and who loses. Subject to limitations imposed by the state of knowledge, economists can apply BCA to answer the efficiency question, although the economist's answers are often controversial, especially when economic efficiency concepts are applied to the environment. The equity question asks whether a choice would distribute its costs and benefits fairly among individuals and segments of society. Economics cannot decide by technical means whether the distribution is fair. It can only expose the distribution of economic consequences. It is then up to those involved in social, political, and legal processes to decide whether that distribution is fair.

In BCA, to the extent possible, all gains and losses that are expected to follow from a particular choice are quantified in monetary terms. The economic definition of value used in BCA is quite specific. It is the amount of money one is willing to exchange for a good or service, broadly defined to include environmental services. Money in and of itself is nothing but a medium of exchange, and a sum of money represents only those things for which it can be exchanged.

This chapter will consider environmental valuation in some detail. Environmental economics is a young field, but no longer in its infancy, and the social value of features of nature has been one of its primary focuses. In the coming pages, we will try to convey the sense of this literature. Before going into environmental valuation more deeply, it is worth pointing out that valuation is only one facet of environmental economics, and the next section will show where valuation fits into the broader context. After that, we will consider in much more depth what is meant by economic value, why environmental economists think environmental valuation is a good idea, and what the limitations of valuation are even within the narrow confines of economics. We will show how valuation is rooted in one branch of Western philosophy, utilitarianism. As the chapter unfolds, other philosophical schools of thought that compete with utilitarianism will be considered. We will eventually show how one's view of the role of valuation in making environmental choices depends on one's philosophical orientation. The chapter ends by drawing some final conclusions together.

3.2 THE MANY ROLES OF ECONOMICS IN ENVIRONMENTAL POLICY ANALYSIS

While most of this chapter will focus on economic valuation, we should make clear at the outset that valuation is only one of the ways that economics can contribute to environmental policy (Hanley et al. 1997; Chapman 1999; Goodstein 2005). Most economic choices get made without having a public decision maker consider benefits and costs at all. Provided certain conditions are met, markets coordinate private choices in ways that promote economic efficiency (Just et al. 1982). Of course, the conditions that lead to satisfactory performance of markets are often not met when it comes to environmental resources. Markets fail to deliver economically efficient outcomes for 2 reasons. First, externalities occur as firms and consumers who do not bear the full costs of their choices have incentives to pollute and do other kinds of environmental harm. Second, many environmental resources such as clean air are public goods, which means, among other things, that when they are provided to one, they are provided to all. Markets do a poor job of providing public goods. Economics has proven useful both in understanding environmental market failures and in suggesting solutions.

Perhaps the most important role that economists play in environmental policy is to help design mechanisms to resolve externalities and assure efficient supplies of public goods in cost-effective ways. For example, cap-and-trade approaches are receiving increasing acceptance. Again consider air pollution as a case in point. Once target levels of emissions have been set at some maximum, emissions permits commensurate with this "cap" can be issued to potential polluters. Allowing trade in permits creates incentives to reduce emissions where it is least costly to do so. An example of tradable permits is the sulfur dioxide (SO_2) allowances issued by the USEPA and traded among electric utilities and on the Chicago Board of Trade (Schmalensee et al. 1998; Stavins 1998). Or, pollution taxes like those levied in Europe on some air pollutants can be used to make private polluters pay the full social cost of their activities, thus giving them market-like incentives to reduce pollution to more efficient levels. Traditional approaches involving technology standards and other so-called command and control measures may achieve the same pollution targets, but are generally more costly (Goodstein 2005). If target levels of pollution can be achieved at a lower cost using economic mechanisms, why not save costs?

Simply expanding the flow of information can facilitate improvements through market mechanisms. If a firm's record of pollution is not known, then other firms or individuals who do business with that firm may be unaware of the environmental effects of the choices being made by the firm in question. Public policies such as right-to-know laws and "green labeling" that require disclosure of harmful practices create marketplace incentives for pollution reduction.

Other economic ideas can be used to design effective environmental regulations. For example, economists can employ analytic methods, such as those from game theory, to predict behavioral responses to different structures.

An additional point that came out in the workshop leading to this book is that economics broadly defined may be an untapped resource for better environmental risk assessments. Suppose a new agricultural pesticide is proposed. Agricultural economists might be asked to predict where and when the pesticide would be used and how much might be applied. Similarly, economics can help predict how future land development patterns are likely to impact a watershed (Erickson et al. 2005).

Furthermore, branches of economics outside of environmental economics may need to be tapped to seek sound policies. For example, 2 of the case studies in this book deal with environmental issues that require addressing the needs of very poor people in developing countries. When the US Navy vacated the Subic Bay area in the Philippines (see Case Study 1), it created an opportunity to establish a national park to preserve several thousands of hectares of mature rain forest, but one of the issues that arose was how to address the needs of local people, including the indigenous Aeta. Proposed surface mining operations in southeastern Madagascar would impinge on the food and fuel resources of local poor people, possibly in positive as well as negative ways (see Case Study 3). Development economics is a well-developed field (Gillis et al. 2001; Todaro and Smith 2003) for addressing such issues, which are not normally studied by environmental economists.

So, there is a lot more to economics than valuation. In fact, valuation might be thought of as the economist's last resort. As we have seen, an economist's first inclination is to seek ways to restructure incentives so that private economic actors can solve environmental problems on their own. Only when such approaches fail to identify effective, politically feasible strategies would most economists turn to valuation.

Nevertheless, in the arena of environmental policy, economists end up falling back on economic valuation quite a lot. Consider again the tradable SO_2 allowances. While this is turning out to be a rather successful mechanism for achieving air emissions targets, what the targets ought to be is unclear without going beyond tradable permits. This is the sort of situation where BCA comes into use. Setting economically justifiable air pollution targets requires knowledge about both the economic benefits and costs of clean air. The remainder of this chapter will look deeper into some issues of central importance to economic valuation.

3.3 ECONOMIC VALUATION FOR ENVIRONMENTAL DECISION MAKING

What is it that economists are seeking to accomplish when they monetize the public's values of environmental amenities? As we have already noted, economists favor markets when markets function well. The problem is that people cannot go to a store and buy clean air for their neighborhoods or unpolluted water in their favorite stream. These are public goods that can be badly affected by externalities. In a sense, what economists try to do is to see how well clean air or unpolluted water or other environmental amenities would stack up economically if markets did work well. Such market-like values can be fed into the public policy arena and compared with more conventional economic values. Armed with this information, public choices can be made that reflect the full economic value of the environmental amenities at issue and not just the values determined by narrow commercial interests. These well-informed choices may occur in the courts, in regulatory agencies, or in legislative arenas.

Valuation studies are conducted within a theoretical framework known as "welfare economics" (Johansson 1991). This branch of economics begins with the so-called Pareto criterion. To see what is involved, suppose a proposed action by some economic actor (e.g., a firm or a government agency) would benefit or harm at least some members of society. That is, the action in question might change conventional economic parameters such as market prices and incomes, but let us also suppose that the quantity or quality of environmental resources would be affected. In such a situation, welfare economics seeks criteria for judging whether the proposed action would make society better off or worse off. The Pareto criterion holds that society would be better off as a result of the proposed action if it would make at least one member of society better off and leave no one worse off.

Although it seems intuitively appealing, the Pareto criterion is not very powerful. Most actions by private economic actors and governments make someone worse off, and thus fail to satisfy the Pareto criterion. Furthermore, the Pareto criterion gives us no yardstick by which to say that some changes make society much better off, while others make it only slightly better off. To be practical, the Pareto criterion must be supplemented.

Economists have — with some misgivings that will be examined momentarily — generally focused on a relaxation of the Pareto criterion called the "compensation test" (Just et al. 1982; Johansson 1991). Suppose that a proposed action will create "winners" who would be made better off and "losers" who would be made worse off. The compensation test states that society will be better off as a result of the action if the winners can fully compensate the losers and still be better off.

In the compensation test, we have the theoretical rationale for valuation (Freeman 2003). Value is defined differently for winners and losers. For winners, the value of the proposed action is their maximum willingness to pay (WTP) to see it put into effect. WTP, added up across all winners, measures the maximum compensation winners would be willing to pay losers. For losers, the value of the change is the minimum compensation that they would require to erase their losses. This is willingness to accept compensation (WTA). The summed WTA of all individual losers is the minimum amount that the winners would have to pay the losers in order to fully compensate their losses. Hence, if the aggregate WTP associated with a proposed action exceeds aggregate WTA for that action, the action would make society better off under the definition of social welfare employed by the compensation test. Where the proposed action would improve the environment, the WTP of those who would enjoy the improvement would be counted in applying the compensation test. Likewise, if the action would lead to environmental degradation, the aggregate WTA of those harmed would be counted among the losses to be borne by losers.

BCA seeks to implement the compensation test (Boardman et al. 1996). The benefits of an action are defined as the estimated WTP of all winners — including environmental winners — who are members of society. Its costs equal the sum of estimated WTA across all the losers — including environmental losers. Hence, to ask whether the benefits exceed the costs is synonymous with asking whether the winners can compensate the losers and still be better off.

As applied in BCA, the compensation test does not require that compensation actually be paid to the losers. Indeed, if compensation were actually paid, there would be no losers and the Pareto criterion would be fully satisfied. The compensation test only requires that full compensation would be possible, not that it would actually occur. For this reason, the compensation test is often referred to as the potential Pareto improvement criterion.

Formal benefit–cost analyses are required by federal law as part of several decision processes, including actions that require regulatory impact analyses (RIAs). For example, since 1981, the USEPA has been required to conduct formal BCAs of all new regulations expected to cost more than $100 million. Even where a full BCA is not required, the compensation test often serves as the basis for valuation. For example, the goal in natural resource damage assessments is to measure the losses resulting from injury to public resources caused by releases of oil or toxic chemicals. Economists would treat those suffering because of resource injuries as losers, and WTA would serve as the conceptual basis for valuing their losses in monetary terms.

Economists recommend including the values of environmental effects with other effects of proposed actions as a way of leveling the playing field. They recognize that the things people normally think of as economic goods and services — TVs, loaves of bread, automobiles, and so on — have economic value. But they also recognize that environmental assets — clean air and water, scenic views, life support services, and the like — are also important to people and have economic values in the form of WTP and WTA that are just as valid and just as relevant to public choices as values for conventional goods and services. When economic actions are under consideration, the values of conventional goods and services will be considered either by private economic actors as they decide what they will do or in the political arena when public entities are making choices. A balanced view of benefits and costs requires that the WTP of environmental winners and the WTA of environmental losers be considered as well.

The compensation test will always be controversial, especially if advocated as a sole decision criterion. Some readers may be surprised to learn that many of the concerns about valuation come from within economics, and in the next 2 sections, we will focus on the theoretical economic issues associated with the compensation test and then the practical problems of actually going out and measuring benefits and costs. Finally, we will broaden the discussion beyond economics by considering some philosophical perspectives on the role, if any, of economic valuation in seeking ethically justifiable actions.

3.4 THEORY-BASED CRITIQUES OF VALUATION FROM WITHIN ECONOMICS

The most hard-hitting criticism from within economics relates to whether choices based on economic values will be fair. A basic tenet built into the compensation test is that a dollar of WTP or WTA for a given winner or loser is the same dollar whether that winner or loser is rich or poor. Hence, a change that would make the rich 1 dollar richer and the poor 99 cents poorer would pass the compensation test. Some economists are therefore quick to point out that there may be good economic reasons to decide against a proposed action even though it passes the compensation test. All that is necessary is for the decision maker to decide that proceeding with the action would be excessively unfair to losers.

Other aspects of the compensation test are also controversial and may limit its applicability in some cases. For instance, the compensation test requires that human preferences be more or less stable or at least predictable over time. If people want one thing today, another tomorrow, and something else the day after, and these changes are not at least somewhat predictable, basing choices on whether today's benefits exceed today's costs would not be a very satisfactory way to proceed, especially where effects are irreversible and affect future generations. If one believes that human preferences are transitory and unpredictable, then the compensation tests would not be of much relevance for most environmental choices.

Concepts of value based on the compensation test are of limited usefulness in cases where affected people live largely outside the dominant market economy. This problem was encountered in trying to measure the losses to native peoples from the Exxon *Valdez* oil spill in 1989. In such cases, indigenous peoples may have very low monetary incomes, produce goods and services in ways that are culturally embedded, and exchange them by gifting and barter rather than in monetary markets. Furthermore, their social and cultural processes for defining, deciding, and expressing their concepts of personal and social good may be quite different from those invoked by the dominant culture.

Crude market measures of loss were obviously inappropriate in assessing the losses of native peoples from the *Valdez* spill — economists could not justify valuing the loss of the walrus hunting (an event of great social and cultural significance) by calculating the cost of purchasing a similar quantity of beef at the supermarket, for example. With such special populations, WTA encounters a particular problem — for cultural reasons, there may be no finite amount of monetary compensation that would have restored Native Americans' sense of well-being or induced them to accept another similar spill. Surrogate measures of damage, for example that use the cost of moving an injured village to a similar but undamaged location, would have been inadequate measures of WTA — they would have ignored the sense of place of people who had occupied their territory for hundreds or perhaps thousands of years and who had, in a sense, evolved culturally in partnership with and as a part of their local environment.

Obviously, the standard economic concepts of value do not travel well across such broad social and cultural distances. Ultimately, as it turned out, the Exxon *Valdez* damage assessment turned to ethnographic methods to supplement the meager and unsatisfactory evidence concerning the economic value of damages to Native Americans in Alaska. Issues of this sort might also arise in any attempt to apply BCA to measure the benefits and costs of national park establishment for the Aeta of the Subic Bay region in the Philippines or of mining on the Malagasy of Madagascar (see Case Studies 1 and 3, respectively).

Other critiques of valuation are rooted in economic philosophy. Economic theories of value define "individual good" as the satisfaction of individual preferences. Technically, the value measures (WTP for gains and WTA for losses) are found through a process that identifies the minimal expenditure that will maintain the individual's baseline "utility" or satisfaction level. Up to this point, there is general agreement among economists.

The philosophical debate begins in attempts to address the nature of social good. The views of most mainstream economists can be linked to the school of philosophy known as "utilitarianism." This is a modern and sophisticated attempt to implement philosopher Jeremy Bentham's idea of "the greatest good for the greatest number." The compensation test and indexes of the standard of living and cost of living are examples of utilitarianism in economics. As we have seen, the framework does not require that losers actually be compensated. Economists who adopt the philosophical view known as "contractarianism" find it intolerable that individuals might be obliged to bear uncompensated harm in service of the public good. Individual consent is emphasized: voluntary exchange of private goods and voluntary taxation for provision of public goods. Accordingly, contractarian positions place great importance on property rights and compensation for individuals who would otherwise be made worse off as the result of actions undertaken for the public good.

These philosophical differences affect how different economists view environmental policy. For utilitarians (most mainstream economists), the work of the economist ends when benefits and costs have been compared and winners and losers described. If decision makers find a policy choice that satisfies the compensation test to be unfair to losers, they are free either to redesign the policy to remedy this problem or to ignore the economic analysis. Contractarians view this whole approach as misguided. An economic case can be made for environmental policies only if the losers actually receive full compensation. According to this group, the fact that the benefits of an environmental policy exceed the costs is no justification at all for adopting that policy. This debate also exists in environmental law, where disagreements continue when environmental policies involve "takings" of private property that require compensation. The debate within economics between utilitarianism and contractarianism continues, with advocates of the latter view seeing a much more limited role for valuation of benefits and costs in making social decisions.

3.5 PRACTICAL PROBLEMS IN MEASURING BENEFITS AND COSTS

A number of other concerns from within economics center on the practical problems of measuring WTP and WTA. In addressing measurement issues, economists distinguish between market and nonmarket values. Loaves of bread, television sets, clothing, and many other commodities that people value are traded in well-functioning markets. Market values for such items convey easily observable information about WTP and WTA. An interesting example of a market environmental benefit arises in the Subic Bay, Philippines (Case Study 1). Giant fruit bats that live in the mature rain forest are important to neighboring farmers as pollinators. Such benefits could be measured using market data on market prices of crops. However, such market signals are not immediately observable for many environmental amenities. In the case of Subic Bay, for example, biodiversity of the rain forests has public goods characteristics that are not captured in the market.

Economists feel most comfortable when dealing with market values. For reasons we have already highlighted, as long as market imperfections are within acceptable limits, market transactions reveal a lot about people's economic values. Market valuation involves what are known as "revealed preference methods," since economic actors reveal their preferences — and hence WTP and WTA — through market transactions.

Nonmarket valuation is more difficult, but progress has been made. In some cases, revealed preference methods have been devised. For example, proximity to a toxic waste site can affect property values as people decide whether to live nearby. If those who buy and sell property in the area understand the risks of living near the site compared to farther away, then lower property values nearer the site reflect property buyers' and sellers' evaluations of the risks. Property values can be compared at various distances from the site to infer WTP for cleanup of those living nearer the site. Of course, if property buyers are unaware or misinformed about the risk, then this approach will not work. In this latter instance, market valuation techniques may not be reliable if the property buyers are unaware of the potential risk.

Revealed preference methods have limitations when it comes to nonmarket valuation. Consider the toxic waste site again. Property values reflect many influences. In addition to risks from toxics, property values across the region may reflect the varying characteristics of the housing; proximity to shopping, medical care, and employment opportunities; the nearness and quality of local schools, parks, and other public facilities; and differences in the willingness of regional residents to accept toxic risks. Hence, it will be necessary to "unbundle" proximity to the toxic site from these other attributes, and that can be a daunting task from a statistical perspective. Furthermore, environmental valuation often involves trying to predict the values of conditions that have yet to be experienced. Suppose, for example, that the economic impacts of locating a toxic waste site in a new area are to be evaluated. Current property values near the proposed site would provide little information.

"Stated preference methods" have been developed to overcome the limitations of revealed preference methods (Mitchell and Carson 1989; Louviere et al. 2000). Stated preference methods query people about their values by posing hypothetical choices in surveys. For example, one might send a survey to people at varying distances from a toxic waste site asking what their WTP would be for cleanup.

Stated preference methods have been controversial. Advocates of revealed preference methods would rather trust what people do in the market system than what they say in surveys. Proponents of stated preference approaches respond that measures of value based on revealed preference might be superior if sufficient data are available. But that is a big "if" for many environmental amenities, and for environmental economists frequently used to stated preference methods as the best or only alternative for estimating WTP and WTA. An overview of the critique and defense of stated preference methods can be found in a special issue of the *Journal of Economic Perspectives* (Diamond and Hausman 1994; Hanemann 1994; Portney 1994).

Research on stated preference and revealed preference methods of environmental valuation continues. Results so far serve to further highlight the limitations on nonmarket valuation. First, when stated preference measures are used, economic researchers have so far had better luck measuring WTP than WTA (Mitchell and Carson 1989). Reliable ways to measure the WTA of losers have yet to be fully developed.

Second, whether stated preference or revealed preference methods are to be applied, it is easier to measure values for small changes than for large ones. Large changes can have far-reaching environmental and economic consequences that are hard to anticipate and measure (Boardman et al. 1996). This problem is particularly prominent in efforts to quantify the benefits of the Clean Air Act and its amendments for the United States as a whole, as summarized in one of the case studies presented later in this volume (Case Study 5). As pointed out there, ecosystem benefits at the national level have been especially elusive.

Third, it is easier to value environmental effects when people are familiar with the resources involved. An economist called in to value the effects of PCB cleanup on local anglers is likely to feel more confident than one who is called in to estimate a national sample's WTP to curb acid rain in New England. The local anglers are likely to be very familiar with the resources affected by PCBs, while residents of other regions of the country may know little or nothing about the resources in New England that are affected by acid rain.

So far, we have stuck closely to modern economics in considering how values are defined, why they might be useful in environmental policy analyses, and what their principal limitations are. It is important to also recognize that noneconomists have weighed in on the efficacy of using monetary values in social decision making.

3.6 CRITIQUES FROM OUTSIDE ECONOMICS: ENDURING VALUE, INTRINSIC VALUE, AND DEEP ECOLOGY

The preceding section showed how even within economics, different philosophical perspectives can lead to very different views about the role of valuation in environmental policy analysis. In fact, this is only the tip of the iceberg. Our goal in this section is to step outside economics altogether and acknowledge how several different moral theories are relevant to policies concerning the environment. We consider these criticisms of valuation in rough order from less to more foundational, beginning with worries that human preferences might be ill considered and ephemeral, and then moving to arguments that value is not really about human preferences, the quest for the social good is not really about maximizing value, and, finally, humans are not the only creatures that matter.

Decisions about ecological risk may commit society to long-lived outcomes (at worst, some natural entity may be lost forever), yet some critics worry that the human preferences that underlie WTP and WTA might be whimsical and ill considered. Preferences may also be limited by what people have experienced, so that they provide little guidance about the value of alternative possibilities. And, preferences may be shaped by the past and embedded in the present, so that they travel poorly into the future. The sense that preferences are impermanent may lead to a quest for some more enduring foundation for value.

More generally, noneconomists sometimes challenge the view of utilitarian economists who refuse any obligation to justify preferences. Economic theory takes individual preferences as its starting point. In theory, preferences are not to be questioned or judged. In practice, economists are not so doctrinaire. For example, economic accounting procedures censor out the value of illicit goods and services. But if some preferences are to be rejected, how are preferences in general to be judged? Economics offers little help. Noneconomists may be ready to argue that some preferences are ill informed, ill considered, and unjustified. An example would be sport utility vehicles, which some consider to be unsafe, polluting gas guzzlers. People who take this point of view may well argue that conservation decisions should be based on natural science, without much consideration of what humans value. Economic rebuttal is constrained by lack of a theory about which human preferences should and should not count.

Going a step farther away from utilitarianism, some would argue that value involves more than individual preferences. Utilitarianism's greatest 19th-century critic, Immanuel Kant (1989), acknowledged a role for preferences, but limited that role mostly to mundane matters like food, warmth, and shelter. Kant insisted that aesthetic judgments, while subjective, involve much more than preference ("I know what I like") — such judgments can make a claim to interpersonal agreement because they can be based on good reasons and shared experiences. The Kantian aesthetic leads to arguments that certain natural entities have intrinsic value — a goodness of their own, independent of human caring (Kiester 1996). There is no consideration in mainstream economics of intrinsic values for the environment or other things. Many environmentalists have views about the values of nature that are closer to those of Kant than those of economists.

Kant also argued that universal moral principles could be found to address the truly important decisions that human and social life requires, an argument that effectively relegates preferences to the issues that are morally inconsequential (at least in a society where food, warmth, and shelter are within everyone's reach). Kant took a duty-based position — there is a duty of obedience to universal moral principles. This perspective leads to a search for moral principles that imply human duties toward natural entities, rather than use of those entities to satisfy human preferences.

The contractarian philosophers believed that the great moral questions are best addressed in terms of rights that must be respected. Contractarian economists espouse a version of this theory, but it is a quite particular version (basically, a strong position that humans have rights in their status quo that must be respected). Rights-based philosophies offer a considerably broader array of positions, some of which have implications for the protection of nature. For example, libertarians might argue that people's rights to enjoy nature oblige other people not to befoul it.

Some believe that environmental choices are not just about humans. Standard economics assumes, along with many strands of philosophy, that humans are the only entities whose preferences count. However, this position has been attacked from many quarters. There are some utilitarians who argue that animal welfare also matters, Kantian aesthetes who argue that natural entities may have intrinsic value, and rights-based ethicists who argue that rights should be extended to other natural entities.

Finally, there are those who believe that environmental choices are not about humans at all. The basic program of deep ecology is to take any or all of the basic moral philosophy approaches and expand the set of entities that matter — that is, entities whose welfare counts, that have rights, and that have a goodness of their own — independent of human concern or patronage.

One idea that many of these viewpoints have in common, and in contrast with the broad sweep of economic thinking, is that natural entities are not always (in the extreme, one might say never) fungible with money. Of course, money is not really the issue — it is, as we have already pointed out, just a convenient metric of value. The real point of contention is substitutability: that is, whether conventional goods and services are (or should be) substitutable for environmental and other natural resources. Nonfungibility arguments are arguments that trade-offs involving natural resources and a bundle of ordinary goods and services are inappropriate, in general or in particular circumstances that can be defined.

By way of pulling together what has been said in this section, the different philosophical perspectives offer insights that some argue should be substituted for economic valuation. Others argue that alternative philosophical perspectives should supplement (and in some cases constrain) economic valuation. Either way, economic valuation does not hold a monopoly on thinking about the role of nature in definitions of the good life and theories of right action.

3.7 SUMMARY AND CONCLUSIONS

We have endeavored in this chapter to show how economic science, in partnership with the environmental sciences, can help environmental decision makers better serve the public interest. We showed that valuation is only one dimension of economics that is relevant to environmental decision making. A principal role of economics in environmental policy is to give policy makers information that will allow them to correct market imperfection through incentives or regulations or by supplying important public goods that might otherwise be unavailable. Much of environmental economics is concerned with restructuring incentives in ways that will lead private economic actors to find desired environmental outcomes on their own.

Having acknowledged that there is much more to economics than valuation, we went on to argue that valuation does have an important role in environmental economics and policy making. Valuation does have a foundation in one of the major schools of Western philosophy, utilitarianism. Furthermore, demonstrating that environmental amenities have economic values and pointing out that progress has been made in developing methods to measure such values can help to level the economic playing field that would otherwise be dominated by narrow private interests.

At the same time, we have acknowledged that valuation has many limitations, both theoretical and practical. Many really tough questions, such as those related to economic fairness or difficult trade-offs that involve nonfungible resources, will require looking beyond economics to find answers. We have suggested that economic valuation can often be useful in environmental policy analysis, but it will not be sufficient in most cases. Those who would make wise environmental choices will need to look beyond economic valuation often in order to discern what is right and good for humankind.

REFERENCES

Boardman AE, Greenberg DH, Vining AR, Weimer DL. 1996. Cost-benefit analysis: concepts and practice. Upper Saddle River (NJ): Prentice-Hall.
Chapman D. 1999. Environmental economics: theory, application, and policy. Reading (MA): Addison-Wesley.
Diamond PA, Hausman JA. 1994. Contingent valuation: is some number better than no number? J Econ Persp 8:45–64.
Erickson JD, Limburg K, Gowdy J, Stainbrook K, Nowosielski A, Polimeni J. 2005. An ecological economic model for integrated scenario analysis: anticipating change in the Hudson River watershed. In: Bruin RJF, Heberling MT, editors, Economics and ecological risk assessment: applications to watershed management. Boca Raton (FL): CRC Press, p 341–370.
Freeman AM III. 2003. The measurement of environmental and resource values. Washington (DC): Resources for the Future.
Gillis M, Radelet S, Snodgrass DR, Roemer R, Snodgrass D. 2001. Economics of development. 5th ed. New York (NY): W.W. Norton.

Goodstein E. 2005. Economics and the environment. 4th ed. Hoboken (NJ): John Wiley.
Hanemann WM. 1994. Valuing the environment through contingent valuation. J Econ Persp 8:19–43.
Hanley N, Shogren JF, White B. 1997. Environmental economics in theory and practice. Oxford (UK): Oxford University Press.
Johansson PO. 1991. An introduction to welfare economics. Cambridge (UK): Cambridge University Press.
Just RE, Hueth DL, Schmitz A. 1982. Applied welfare economics and public policy. Englewood Cliffs (NJ): Prentice-Hall.
Kant I. 1989. Foundations of the metaphysics of morals. 2nd ed. Beck L, trans. Englewood Cliffs (NJ): Prentice Hall
Kiester R. 1996. Aesthetics of biological diversity. Hum Ecol Rev 3:151–157.
Louviere JJ, Hensher DA, Swait JD. 2000. Stated-choice methods: analysis and application. Cambridge (UK): Cambridge University Press.
Mitchell RC, Carson RT. 1989. Using surveys to value public goods: the contingent valuation method. Washington (DC): Resources for the Future.
Portney PR. 1994. The contingent valuation debate: why economists should care. J Econ Persp 8:3–17.
Schmalensee R, Joskow PL, Ellerman AD, Montero JP, Bailey EM. 1998. An interim evaluation of sulphur dioxide emissions trading. J Econ Persp 12:53–68.
Stavins RN. 1998. What can we learn from the grand policy experiment? Lessons from the SO_2 allowance trading. J Econ Persp 12:69–88.
Todaro MP, Smith SC. 2003. Economic development. 8th ed. Redding (MA): Addison Wesley.

4 Valuation Methods

W.L. (Vic) Adamowicz, David Chapman, Gene Mancini, Wayne R. Munns, Jr., Andrew Stirling, and Ted Tomasi

CONTENTS

- 4.1 Introduction 60
- 4.2 "Open" and "Closed" Approaches to Appraisal 61
- 4.3 A Description of Valuation Techniques 62
  - 4.3.1 Stated Preference Methods 63
    - 4.3.1.1 Contingent Valuation 63
    - 4.3.1.2 Choice Experiments or Attribute-Based Methods 64
    - 4.3.1.3 Environmental Damage Schedules 65
    - 4.3.1.4 Variants of Stated Preference Elicitation Methods 65
  - 4.3.2 Revealed Preference Methods 65
  - 4.3.3 Replacement Cost and Opportunity Cost Methods 66
  - 4.3.4 Economic Impact Analysis (EIA) 67
  - 4.3.5 Habitat Equivalency Analysis (HEA) 67
  - 4.3.6 Energy Methods 69
  - 4.3.7 Environmental and Sociocultural Indices 69
    - 4.3.7.1 Green Accounting 70
    - 4.3.7.2 Ecological Footprint 70
    - 4.3.7.3 Mass Balance and Natural Step 70
    - 4.3.7.4 Genuine Progress Indicator 71
    - 4.3.7.5 Environmental Sustainability Index 71
  - 4.3.8 Technology Assessment 72
  - 4.3.9 Life Cycle Analysis 72
  - 4.3.10 Social Impact Assessment 73
- 4.4 Evaluation Criteria 73
  - 4.4.1 The Ability to Capture Heterogeneity 74
  - 4.4.2 The Ability to Incorporate Complexity of Natural Systems 77
  - 4.4.3 The Handling of Incomplete Knowledge in Decision Making ...78
  - 4.4.4 Ordinal versus Cardinal Measures 81
  - 4.4.5 Responsiveness to Context 82
  - 4.4.6 Practicality 83
- 4.5 Plug and Play Tools: Evaluation Aids That Can Be Applied to All Methods 84
- 4.6 Conclusions 85
- References 86

Appendix 4.1: Glossary...91
Appendix 4.2: Decision Analysis Tools...92
A.1   Decision Analytic Frameworks 93
     A.1.1   Conceptual Approaches 93
          A.1.1.1   Stochastic Dynamic Programming..................................93
          A.1.1.2   Bayesian Analysis 94
          A.1.1.3   Multicriteria Analysis 94
          A.1.1.4   Simulation Analysis...94
     A.1.2   Applications of Decision Analytic Frameworks..............................94
          A.1.2.1   Soft Systems Methodologies..95
          A.1.2.2   Adaptive Methodology for Ecosystem Sustainability
                    and Health (AMESH)..95
          A.1.2.3   Ecosystem Management Decision Support (EMDS).........96

4.1 INTRODUCTION

This chapter will describe and then assess the usefulness of various methods for the valuation of environmental resources. To count as a valuation method, the approach must aggregate in some fashion across at least 2 indicators of different environmental services. Thus, an index of habitat suitability for a particular species or a species diversity index, while embodying several aspects of ecosystem structure or function, essentially is a descriptive measure of a single ecosystem service, and is not itself a valuation method by our definition. However, an index that is a function of several services, such as primary production and a diversity index, could be a valuation function, since it embodies trade-offs among services, and trade-offs among services is our underlying definition of value, as discussed in this chapter and Chapter 3.

We note that the idea of value must be related to the decision maker at issue. All biological organisms, for instance, must respond to "value" signals based on net energy. Thus, a fish could be said to value alternative prey according to the net energy of eating it, that is, energy intake net of energy expended. For people, economists generally model values based on individual preference trade-offs. Higher level decision makers may respond to additional values, and also recognize that lower level entities can be modeled (and their behavior predicted) based on value cues at those lower levels. Thus, the higher level decision maker may recognize that people respond to preference-based value incentives, and ecosystem changes can be modeled based on energy, in predicting alternative outcomes, which then are valued according to the decision maker's value system. Using this approach, energy (or emergy or exergy) analysis, which ranks both human and ecological outcomes by net energy flows, is not a human value-free system since it implies that decisions should be made according to energy measures.

We include both economic and noneconomic value measures. To count as an economic valuation, the method must appeal to trade-offs or other ranking relationships expressed according to individual preferences.[1] But such trades need not be

---
[1] We include here preferences that may not be continuous, and so lexicographic preferences would be included.

Valuation Methods

denominated in monetary terms. Dollars are but one potential "unit of account," and we may instead estimate the willingness of individuals to trade one environmental good or service directly for another, or for some other good. Should the method denominate the trade in a currency such as dollars, we shall call the method a "monetary valuation." In this case, the trade involves the environmental good or service for a generalized basket of other market goods, rather than some specific other good.

If a market exists for the good or service in question, then the value of that service is readily measured in dollars. The market is a convenient source of data on trades that are summarized in a dollar price. We note that environmental services could be monetized if institutions were designed to establish markets. These institutions are evolving rapidly. For example, the use of tradable emission permits is expanding, along with the attendant markets and observable prices.

The tools examined are meant to inform some type of decision process. The exact operation of the decision process per se is not in this chapter's purview. Thus, numerous ways to aggregate individual preferences into some form of group preference (e.g., voting schemes, or consensus-building processes) are not addressed here. Similarly, most economic indices, including benefit–cost analysis (BCA), have been proposed explicitly as inputs to a decision process, not as the sole criterion sufficient to base a decision on (for more on BCA, see Chapter 3). We do, however, review various conceptual approaches to decision analysis in Appendix 4.2 at the end of this chapter. These involve various ways of organizing and summarizing data.

The goal of the chapter is to evaluate the various methods of valuation according to several criteria discussed below. An overarching issue is the ability of the method to improve environmental decision making. Obviously, this depends on the decision context. Thus, we are not asking whether tools are good or bad, but rather seek to clarify the decision contexts in which the tool will be most (or least) useful. In particular, we address the question of when monetization is useful, and when it is not.

4.2 "OPEN" AND "CLOSED" APPROACHES TO APPRAISAL

Before delving into the task of describing and assessing valuation approaches, we discuss a more general issue associated with open and closed approaches to appraisal. We begin our chapter with this discussion because it permeates all methods of valuation discussed below. In this context, "open" simply means employing methods and approaches that incorporate external views or inputs, fostering transparency as it were. In contrast, "closed" may not specifically incorporate external views or inputs, and it is likely that there could be some elements of the approach that are not well-known, tested, or approved outside of the individuals or groups employing them (Stirling 2005).

Risk assessment, benefit–cost analysis, and other approaches to health, environmental, and wider impact assessment are conventionally conducted as a means to provide concrete prescriptive policy recommendations. It is well understood that such methods can only offer aids, rather than substitutes, for decision making.

However, there exist strong and understandable institutional pressures to present results in a form that is as concrete and useful as possible for decision makers. Even where there exists no institutional pressures in relation to substantiating prior positions concerning the approval or restriction of particular options, decision-making processes typically value technical analysis as a firm basis for justifying policy making in the face of inevitable political contention and challenge. Even though the analysts themselves are often extremely well aware of the crucial roles played in their analysis by uncertainty, ambiguity, and ignorance, these are not always fully conveyed to decision makers or wider policy-making debates.

It is for this reason that technical analysis is often conducted in a relatively closed fashion, emphasizing risk over uncertainty, underemphasizing the implications of different framing assumptions, and neglecting to discuss the possible importance of unknowns. Interestingly, the same observation might be made of the way that processes of qualitative deliberation and inclusive participation tend to drive for consensus, understating uncertainty and ambiguity. It is here that the transparency and auditability of quantitative techniques come to the fore. Although applicable in principle under a wide variety of such appraisal tools, techniques like sensitivity and scenario analysis remain relatively underused. Under a more open approach — by contrast — such techniques offer (in conjunction with appropriate participatory processes) a means to be as rigorous about the handling of framing assumptions in analysis as about the treatment of the data themselves.

4.3 A DESCRIPTION OF VALUATION TECHNIQUES

In this section, we describe a number of techniques that could be used to evaluate the outcomes of policies or projects. We do not limit ourselves to techniques that are economic in nature or that employ monetary metrics in valuation. We have deliberately chosen to cast the net broadly to include techniques from biology, economics, decision research, and other areas. We provide brief overviews of these techniques, recognizing that entire volumes can be written about each technique. In the next section, we present criteria that we use to evaluate these techniques, and proceed to develop a matrix that presents our evaluation.

Before embarking on the remaining task, a comment about the relationship between the valuation of flows of ecological (or economic) services and the valuation of changes in stocks of ecological (or human-made capital) assets is in order. Generally, a capital asset, whether ecological or human made, is measured as a stock, such as the standing biomass in a forest or a biomass of fish. The value of an asset is a derived value, derived from the flows of services it can produce. Thus, a forest can produce flows of aesthetic or recreational services, wildlife habitat services, and timber supply. The change in the value of a stock is then computed from the change in the value of the service flows that changes in the stock produce. Our chapter primarily looks at methods used to value service flows. However, this does not limit the ability to infer stock changes, and some of the methods do have explicit recognition of stocks (such as the green GDP, or gross domestic product).

Valuation Methods

4.3.1 STATED PREFERENCE METHODS

Stated preference (SP) methods are a very broad class of methods that elicit preferences directly from people.[2] These preferences are listed as "stated" (as compared to "revealed") because there are no behaviors being observed that can be used to identify the preferences. That is, the preferences are elicited from conversations with individuals rather than from purchases, votes, expenditures of time in activities, or other observable behaviors. Stated preference methods are implemented in many forms, such as surveys of the general public, small-group discussions, and deliberative processes. They also use very different preference elicitation tools. Individuals are sometimes asked to choose between options presented to them, to rate or rank options, or to identify the best item on a list and contrast it with the worst. Stated preference methods are also framed in various ways. Some are framed as if they were votes, some as purchases, and some as specific decisions that a company, government, or household would undertake. Stated preference methods are generally consistent with economic valuation since they elicit preferences from individuals, and some stated preference methods provide monetary values. The listing below provides a brief description of some of the stated preference methods in use today.

4.3.1.1 Contingent Valuation

Contingent valuation asks individuals to express preferences by identifying their willingness to trade money for a well-defined policy or project. Individuals are asked to value the change induced by the project (e.g., change in environmental quality or risk). The most popular current approach involves describing a referendum in which individuals vote on whether or not they would accept the change in quality and the payment. Other variants include eliciting specific monetary values by either presenting people with a card with various categories of amounts they would be willing to pay (including zero) or asking people to identify that amount directly. When used in policy analysis, representative sampling approaches would be used and the surveys would be conducted using in-person interviews, the Internet, or the mail. The distinguishing features of contingent valuation are the focus on monetary valuation and the detailed characterization of the good to be valued (e.g., environmental quality change).

Issues that arise in the development of contingent valuation surveys include the communication of the good and the changes involved, the method for eliciting payment, and the degree to which the survey is treated as "consequential" (interpreted as the respondent's feeling that he or she may actually have to pay these amounts and

---

[2] A significant concern arising in the economics literature is the degree to which the valuation methods suffer from strategic behavior. This is a form of response behavior in which the individual being asked about value does not reveal a true value but provides a value that he or she feels will enhance his or her utility or well-being. While strategic behavior can occur in stated preference (hypothetical) situations and revealed preference cases, there is more concern about such behavior in stated preference approaches. Concern about strategic behavior could be an evaluation criterion, but this can make things rather complicated. There is an argument, for instance, that valuation is open to strategic behavior on the part of analysts and sponsors in the choice and framing of methodologies, as well as by subjects in stated and revealed preference methods. We therefore do not include strategic behavior in our list of evaluation criteria. Nevertheless, decision makers should recognize the potential for stated preference — and other — methods to be subject to strategic behavior.

that his or her response will influence public policy). The latter is an important component in determining if the survey is incentive compatible (i.e., providing an environment where truthfully revealing preferences is the respondent's best strategy). This issue is also related to the efforts taken to reduce the hypothetical nature of the survey. Other key development issues include the statistical design of the elicitation approach, and the analysis of the responses. Since this technique is administered to representative samples of individuals, it can be used to assess the distribution of benefits over the population and the systematic factors that affect that distribution. Readings that provide overviews of contingent valuation include Carson (2000) and Boyle (2003).

4.3.1.2 Choice Experiments or Attribute-Based Methods

Choice experiments or attribute-based methods are similar to contingent valuation methods in that they ask individuals to state their choice from alternatives and use these statements of choice to identify preferences and values. However, choice experiments characterize the goods in terms of collections of attributes rather than as a single description. This is useful when the situation or good is perceived by the respondent as a collection of attributes and when the valuation of attributes, separately from the entire policy, is desirable. The theoretical basis for choice experiments is a fusion of Lancaster's theory of utility from attributes (Lancaster 1966) and McFadden's development of random utility theory (McFadden 1974). This approach has been used in recreation valuation, for example, to help identify how changes in one attribute of the experience (fish catch rate) are valued holding other attributes (scenery, number of picnic sites, etc.) constant.

Choice experiments often present individuals with many sets of choices where the attribute combinations vary according to an experimental design. This experimental design helps identify the contribution of the attributes to the overall utility of value. Choice experiments can be presented as purchases ("Which of these options would you purchase?"), votes, or other elicitation approaches. The sets of options used in choice experiments can also be used in ranking exercises, or individuals can be asked to rate each of the options on some rating scale. Combinations of these approaches have also been used.

Development issues in choice experiments are similar to those in contingent valuation: communication and framing, consequentiality and incentive compatibility, experimental design, and analysis. Choice experiments are also commonly based on representative samples of individuals, which allows for the assessment of the distribution of preferences and benefits. Descriptions of choice experiments include Adamowicz et al. (1998) and Holmes and Adamowicz (2003).

Both contingent valuation and choice experiments provide estimates of monetary value for changes in environmental quality (or other goods characterized in the valuation context). Choice experiments have the ability to use any one of the attributes as numeraire. This allows choice experiments to provide measures of trade-off weights between any attributes that are part of the valuation frame. In theory, contingent valuation could also be used to identify such trade-offs; however, it has not been used in such a fashion.

Valuation Methods

4.3.1.3 Environmental Damage Schedules

A variant of choice experiments (a simplification in some ways) is the method of paired comparisons. This method provides individuals with pairs (e.g., pairs of public goods) and asks them which they prefer. This technique provides a ranking of goods (public goods, private goods, and perhaps money) which can be used to develop damage schedules. Damage schedules provide a list of environmental damages in order of their perceived harm to the environment. This ranking of harm can be used to develop corresponding monetary values for the damage, similar to the schedules that insurance companies provide for benefits provided for human injuries or death. The environmental damage schedule approach has not been implemented in any significant policy, and there is some debate about the methods used to develop such schedules. Since these schedules are based on preferences elicited from groups of respondents, they could be based on representative samples and the measure could be analyzed to assess heterogeneity over respondents. However, the idea of a fixed damage schedule seems to preclude considerations of heterogeneous preferences. An overview of environmental damage schedules can be found in Chuenpagdee et al. (2001), while a review of pair comparison approaches is presented in Brown and Peterson (2003).

4.3.1.4 Variants of Stated Preference Elicitation Methods

The sections above have outlined the main approaches used in stated preference elicitation and valuation in economics. These approaches are primarily based on stated choices, as choices correspond well with economic theory, particularly random utility theory and economic demand theories based on attributes. Alternative elicitation approaches include methods that ask respondents to rate or rank alternatives or in some other way describe a preference ordering over goods or bundles of goods. These approaches introduce some additional complexities or assumptions into the analysis in order to be consistent with economic theory. However, they are in principle consistent with the notion of preference elicitation in an economic context. More importantly, there are variations on the process within which these methods are implemented. Within economic valuation circles, representative sampling in person, via the Internet, or via mail techniques seems most popular. These approaches, however, are somewhat distant from the respondents and do not encourage much discourse or learning about the issue. Alternatives that include deliberation, discussion, and debate have been developed and implemented. The major difference in these approaches is that they are changing the context of the decision making by incorporating social interaction and learning. The trade-off is that these approaches tend to operate with smaller samples of individuals that are not usually representative of the population. The analysis of heterogeneity in such cases is also somewhat limited. Multicriteria analysis is an example of one of these stated preference techniques that typically involves smaller samples of individuals under uncertainty (see Appendix A.1.1.3).

4.3.2 REVEALED PREFERENCE METHODS

Revealed preference (RP) methods (or, sometimes, "indirect methods") are economic valuation techniques in which individual value trade-offs are inferred from data on choices made regarding goods with a known linkage to environmental goods. For environmental goods, such choice data can come from market or nonmarket settings.

The hedonic method examines market prices for consumer goods, real estate, and/or wage rates for jobs. These prices depend in part on the environmental or safety attributes of alternative goods, houses, or jobs. Food that is certified safe or lumber that is eco-labeled is linked to pesticide use or timber management regimes. Goods linked to more highly valued environmental attributes sell at a premium. This premium may be used to infer a value for health risk or sustainable timber harvest. A property location is linked to environmental attributes of the location. Waterfront properties on clean water sell for more than those on polluted waterways, and these types of data can be used to infer the value of clean water. Just as house prices depend on house and lot features as well as neighborhood and school quality, house prices depend too on air quality, noise, and so on. Wages in desirable locations are higher than those in less desired locations, all other job attributes held fixed. Hedonic methods have been used to value health risk, clean air, clean water, contaminated sites, airport or highway noise, and other environmental goods. Good descriptions on the hedonic method include Freeman (2003) and Taylor (2003).

Another set of prices that can be used in this fashion is stock prices. This might be done from the standpoint of the way in which both attributes and behaviors of a corporation (e.g., environmental releases, and social aspects of the firm) affect stock prices.

The travel cost method is similar to the hedonic technique, but oriented to recreation choices. The data collected are not on market purchases, but rather surveys are used to ask how often and where people recreate. The cost of getting to a site (hence, the name) can be computed, and plays the role of a price. The willingness to spend travel and time costs to access different recreation sites can then be related statistically to data on the attributes of those sites. Example attributes include harvest rates for fishing or hunting, water quality, beach width, educational opportunities, and aesthetic measures, among others (Phaneuf and Smith 2006).

With experimental data, the analyst constructs a choice situation in a "laboratory" setting and recruits individuals to the experiment. Subjects are given real goods or money (or a token that can later be traded for money or a good) and make choices about real but constructed events. The experimentalist can alter the types of goods at issue, information, terms of interaction among subjects, and other features of the choice setting. The choices made by the experimental subjects can then be used to infer value, just as with hedonic data in a market setting. Examples include killing a sapling (Boyce et al. 1992) or planting trees in Atlanta (Boyle et al. 2004).

4.3.3 REPLACEMENT COST AND OPPORTUNITY COST METHODS

The cost of replacing an environmental service if it is lost is often used as a measure of value. For example, wetland services lost to development may be valued by the cost of creating replacement wetlands. Similarly, an endangered species might be valued via the cost of habitat protection. These costs are opportunity costs, that is, what could have been done with the resources expended in building the replacement wetland, or the foregone value produced by the habitat if it is not protected. As a

Valuation Methods

general matter, these cost measures confuse cost with benefit, and are not a legitimate value measure. The whole purpose of the valuation exercise is to determine whether the replacement or mitigation action is a "good idea" or "worth the cost," and so to use cost as the measure of value begs the central question.

That said, cost provides a hurdle that benefit must exceed if efficiency is to be enhanced by a decision, and sometimes exceeding that hurdle is obvious. For example, consider the CALFED case study (Case Study 6). To keep water in the Sacramento River Delta to provide fish habitat required denying water at times to southern California. The cost of water for fish habitat is the amount of money southern California was willing to accept to make them as well off as with the water delivered. The withholding of water passes a benefit–cost test (and is an actual Pareto improvement, not just a potential one — see Chapter 3) if saving the salmon is worth more than the cost of protection. This might be obvious and not in need of careful measurement. Indeed, the US Endangered Species Act (ESA) may be construed as a social determination that avoidance of species extinction generally is very highly valued, and most likely will exceed any costs imposed by the program. Naturally, since not all social decisions automatically and correctly embody social values (if they did, valuation would not be needed), the economic efficiency of ESA may still be open to question. Of course, as the salmon recovered and stocks grew away from the critical level, the value question became more ambiguous.

4.3.4 ECONOMIC IMPACT ANALYSIS (EIA)

Economic impact analysis (EIA) comprises a set of tools that calculate the effect of alternative actions on the level and composition of economic activity in a region. These tools trace economic interactions in terms of the amount of money expended by economic sector, and other indicators such as numbers of jobs and net migration. The computed indicators are traditional measures of gross regional product, income, sales, employment, and the like. There are varieties of implementation methods that differ in degree of sophistication and flexibility (references to IMPLAN [Minnesota IMPLAN Group] and REMI [Regional Economic Models, Inc.]) that typically are based on input–output models (Miller and Blair 1985). But all EIA methods focus fundamentally on expenditures and hence are unlike benefit–cost measures, which are based on WTP. Since much of the change in economic activity measured as expenditures represents a transfer from one economic entity to another, these (for the most part) "drop out" of the benefit–cost measure, as they do not represent net gains or losses. The benefit–cost measures deduct expenditures from gross WTP to arrive at consumer and producer surplus measures of value. Nevertheless, EIA provides important information on employment and output impacts at a regional level.

4.3.5 HABITAT EQUIVALENCY ANALYSIS (HEA)

Habitat Equivalency Analysis (HEA) and the subsequently developed Resource Equivalency Analysis (REA) are methods developed in the early 1990s to evaluate the amount of environmental improvement or creation necessary to offset a specific environmental loss (Unsworth and Bishop 1994; Chapman et al. 1998). The methods are based on the concept that natural resources provide a stock and flow of services often called "natural capital" and "natural income." While the underlying foundations of HEA and REA are in principle economic, this discussion will focus on the specific set of circumstances where the analysis can be reduced to a comparison of ecological measures. The ultimate output of HEA is measured in a quantity of environmental improvements (restoration).

There are numerous applications of HEA and REA, including natural resource damage assessments, Section 404 of the US Clean Water Act mitigation requirements, Section 316(a) and (b) restoration actions, and environmental impact of policy assessments (Chapman et al. 1998; Penn and Tomasi 2002). The concept of HEA has also been applied to inform decisions about the level of cleanup appropriate at a specific site. In this manifestation, it is called Net Environmental Benefit Assessment (NEBA; Nicolette et al. 2001).

HEA can be described through 3 basic steps: choice of metric, evaluation of loss and gains, and scaling. The choice of metric is critical to the entire analysis. The metric chosen should adequately represent the types of environmental changes that occur due to *both* the injury and to restoration projects. The choice, or construction, of the metric, will in large part be a function of the complexity of the system being evaluated and how adequately a simple measure can capture both the losses and gains. The choice of metric is dependent upon both the complexity of the systems being analyzed for losses and gains and the mechanisms of environmental change. If one is looking at physical injury to seagrasses, perhaps a measure of short shoot density will be an adequate metric to describe both the loss due to a vessel grounding and the benefits of restoration actions. However, in a system where sublethal effects from metals and organic chemicals among multiple species are of concern, and common restoration actions will try to address these multiple species, a more complex metric that might be some weighted average of multiple measures would be necessary.

A calculation of the environmental losses and gains from specific actions can be developed through a mapping of the metric onto 4 conditions: the condition of the resources with the environmental loss (injury); the condition of the resources without the loss (baseline); the condition of the restoration site(s) before any actions are undertaken (restoration baseline); and, finally, the condition of the restoration site(s) after actions are implemented (restoration benefits). When the same metric is used to evaluate each of these cases, the relative losses and gains are comparable. This step includes a comparison of the time periods over which gains and losses occur through the use of discounting, and incorporation of the rates of recovery of the injured site(s) and restoration site(s).

Scaling is the final step used to describe the calculation to determine the size or quantity of restoration projects necessary such that the quantity of environmental gains is equal to the estimated environmental loss. This step will result in a recommendation on the quantity of specific restoration actions necessary to just offset the service reductions of concern.

NEBA applications of HEA follow the first 2 steps above; however, the final focus is on comparing alternative actions that affect the environment. An example would be to compare the net environmental benefits of alternative levels of sediment remediation. This approach would be used to compare the benefits to the environment from increasing levels of sediment remediation, say, by dredging increased acreage. However, as one dredges more area around a hot spot, the level of net service improvement decreases. NEBA can be used to identify the area where benefits no longer outweigh the injury from the contaminated sediments.

4.3.6 Energy Methods

The concept of energy as a biophysical measure of value has its foundation in the maximum power principle of Lotka (1922). Simply put, that principle suggests that self-organizing systems, such as ecosystems and economies, tend toward maximization of energy utilization. Methods that utilize measured and modeled energy to define the value of environmental system components include emergy and exergy. "Emergy" is defined as "the available energy of one kind previously required to be used up directly and indirectly to make a product or service" (Odum 1996). Emergy is quantified in terms of emjoules, which can be calculated for goods and services by relating the energy content of the entity back to the solar joules ultimately used to create the good or service. It has been argued that emergy measures real wealth, independent of the trade-offs people make (Odum 1996). Conceptually, it provides a means to value environmental system components and processes based upon the organization of the system itself. To support analyses of economic systems, emergy can be translated to "emdollars" by dividing emergy by the ratio of total emergy within a system to the total money in the economy. Emergy or emdollars can be used to calculate a number of indices to support analysis of the condition of economies and ecosystems. The emergy approach has been used to evaluate a number of systems, including shrimp mariculture (Odum and Arding 1989), electrical power (Odum 2000), and the cycling of gases in the biosphere (Odum 2000).

An emergy evaluation consists of describing an environmental system in terms of its components (storages), flows among components, and flows into (inputs) and out of (outputs) the system. Typically, this description takes the form of an energy systems diagram (Odum 1971), supported by the specification of system component emergies and the mathematical equations describing flows among components. Emergies can be calculated in a number of ways, as summarized in Odum (1996).

For a number of reasons, the emergy concept has failed to gain traction and wide acceptance in policy and decision making. Recently, Tschirhart (2000) proposed an approach to integrate general equilibrium models for ecosystems and economies, using energy maximization (net energy intake) by organisms as the organizing principle for the dynamics of species' interactions. Conceptually, equilibria occur in the ecosystem model when the market demands by predators for each prey species' biomass are equal to the supply of the prey species' biomass at some set of energy prices (the cost of capturing biomass as prey). Integration of ecosystem models with economic models is accomplished through incorporation of the population sizes of species as variables in the economic model, allowing the choices people make to affect those sizes.

4.3.7 Environmental and Sociocultural Indices

A variety of indices have been developed and used in recent years that attempt to integrate an array of sociocultural and environmental elements into broad-scale economic or sustainability measurements. A few of the most prominent indices include green accounting, or green GDP, which may reflect a variety of green accounting principles; the ecological footprint (EF) concept; the Genuine Progress Indicator (GPI); the Environmental Sustainability Index (ESI) and the parallel Environmental Performance Index (EPI); and mass balance, or natural step. Each of these analytical constructs aggregates multiple underlying assumptions and, with the exception of the EF, aggregates multiple metrics in an attempt to arrive at economic or sustainability representations that are more comprehensive or informative than gross national product (GNP) or gross domestic product (GDP). Each of these indices will be briefly described and evaluated in a comparative matrix with other resource valuation techniques and methodologies.

4.3.7.1 Green Accounting

Green accounting focuses on the recognition of and accounting for environmental costs associated with natural resource depletion and impact in addition to the economic benefits derived from natural resource extraction and use. It explicitly calls for a full integration of such benefits and costs in measures of economic health and performance, such as GDP. The broad principles of green accounting incorporate concepts such as natural capital and sustainability as presented in Jansson et al. (1994). The calculation of green GDP is intended to account fully for the relevant benefits and costs of natural resource use and is anticipated to be more comprehensive and informative than the conventional GDP. The Genuine Progress Indicator described below is in the spirit of such green accounting indices, but relies less on rigorous economic aggregation issues.

4.3.7.2 Ecological Footprint

The ecological footprint can be defined for a given human population as the total area of terrestrial and aquatic ecosystems required to produce the resources consumed by that population, including the assimilation of its generated wastes (Rees 1992, 2000). The EF has appeal as an indicator to summarize and represent total human resource use that is easily communicated. Additionally, the EF has been proposed as a sort of sustainability guideline where a population's EF is compared to the total area under its control. Where the EF exceeds the area of control, resource use may not be sustainable.

The EF is an aggregated index in that it embeds all aspects and assumptions about resource use and demand in a single areal unit. Despite the EF concept's calculational simplicity, it has generated considerable deliberation and debate and has resulted in a dedicated special issue of *Ecological Economics* (March 2000, Vol. 32, Issue 3). Costanza (2000) and others have criticized the concept as unable to characterize or fully represent its significant underlying assumptions (e.g., technological constraints on society), variabilities, and complexities.

4.3.7.3 Mass Balance and Natural Step

The concept of mass balance is based on the fundamental physical principle that, whatever other changes are involved, matter itself is neither created nor destroyed

Valuation Methods

in normal industrial processes. Therefore, the mass of inputs to a process, industry, or region balances the mass of outputs as products, emissions, and wastes, plus any change in stocks. The analysis of mass balance thus provides one way to evaluate contrasting distributions and relative efficiencies in flows of useful material resources (FFF 2003). Of course, there are corresponding challenges in linking these factors to other aspects of value.

One particular approach that builds on these foundations as a framework for the appraisal of sustainability in organizations, industries, and nations is known as the "natural step." Promulgated as a proprietary approach tailored especially to use in industry, this is based on 4 basic "conditions for sustainability" (Natural Step 2003). The first 2 are related to mass balance concepts, in that they propose a halt to systematic increases in environmental concentrations of substances extracted from the Earth's crust and substances produced by society. The third condition requires a halt to the systematic physical degradation of living resources, such as forests. The 4th condition requires the meeting of human needs. This provides one potentially useful general framework for the development of more tailored minimal criteria for sustainability. However, it offers less in addressing key valuation issues concerning the characterizing, prioritizing, and trading of contending metrics of value.

4.3.7.4 Genuine Progress Indicator

The Genuine Progress Indicator is an annually revised aggregated index, which is calculated somewhat along the lines of general green accounting principles. The GPI accounts for uncompensated contributions to the economy as credits, while it debits factors such as pollution and crime (Cobb et al. 2001). While the GPI is developed within the same framework as the GDP, it explicitly accounts for a number of sociocultural and environmental benefits and costs not included in GDP calculations. The GPI includes personal consumption, of course, but also includes an array of economic, social, and environmental costs such as unequal income distribution and underemployment, family breakdown, and depletion of nonrenewable resources, while adding benefits such as parenting and volunteer work. A total of 25 categories are included in the GPI calculation. In 2000, US per capita GDP was $33 497, while the calculated per capita GPI was calculated as $9550; the longer term trend for the GPI (1950 to current) is substantially and consistently lower than the GDP.

Application of the GPI has been attempted at the state level (Minnesota Planning Environmental Quality Board 2000), where it was identified as a potentially useful progress indicator. It was also criticized, however, as incorporating problematic monetary value assumptions (e.g., wage rates for household workers and wetland elimination costs), for variability of estimated costs, and for the general lack of regional or local data for application below the national level.

4.3.7.5 Environmental Sustainability Index

The Environmental Sustainability Index is an evolving methodology that uses 20 indicators and subsets of variables to measure progress toward environmental sustainability for 142 countries (SEDAC 2004). Each indicator combines 2 to 8 variables to generate 68 data sets to evaluate 5 core elements on a national basis. The core elements include global stewardship, social and institutional capacity, environmental systems, reducing human vulnerabilities, and reducing stresses. A parallel Environmental Performance Index ranks countries based upon various air and water quality, climate change prevention, and land protection metrics.

Integrated environmental indicators in the ESI include such variables as concentrations of oxides of sulfur and nitrogen in air, biochemical oxygen demand (BOD) and phosphorous in water, carbon economic efficiency (CO_2 emissions per GDP), and glass and paper recycling rates. The relevant data sets are aggregated across the ESI core components to generate a single score, in 2002 ranging from 73.9 (Finland) to 23.9 (Kuwait). Higher scores are intended to indicate greater national sustainability, and the annually revised scores are anticipated to document trends in addition to allowing cross-national comparisons.

4.3.8 TECHNOLOGY ASSESSMENT

Technology assessment refers to a family of methods focusing on the appraisal of positive and negative effects of technical artifacts, technologies, or technological systems. Examples might be alternative urban transportation systems based on the automobile compared with the bus or the urban light railway, or alternative automotive fuel infrastructures such as biofuels or compressed natural gas for internal combustion engines, electricity for batteries, or hydrogen for fuel cells. Attention typically extends to a wide range of effects, including human health, environmental impacts, economic benefits, and social and cultural implications. As such, technology assessment can provide a framework for the application of a number of other methods considered elsewhere in this chapter, for instance being used in conjunction with multicriteria or life cycle analysis. In this regard, the distinguishing feature of technology assessment as a framework is that it focuses on the dynamics of technological innovation, addressing a range of technologically defined alternatives, and implies attention to innovation and production systems taken as a whole (Loveridge 1996). Technology assessment is usually applied ex ante, in anticipation of a range of alternative technological trajectories. There exists a large variety of specific approaches, involving various matrix frameworks and accounting and aggregation protocols, and, more recently, the advent of "constructive," "interactive," and "participatory" technology assessment procedures that concentrate on ways to elicit pertinent information and relevant values from a wide range of social constituencies (Rip et al. 1996; Grin et al. 1997).

4.3.9 LIFE CYCLE ANALYSIS

Life cycle analysis refers to a framework for the appraisal of alternative products, production processes, or infrastructure investments, which focuses particular attention on the challenges associated with defining the boundaries of the industrial, technical, or policy systems under scrutiny (van den Berg et al. 1995). Rather than looking at positive effects or broader social and economic issues, life cycle analysis usually restricts attention to the negative environmental or health impacts. A number of procedures have been developed for systematically tracking the magnitude of the

Valuation Methods

impacts associated with the full resource chains and facility "life cycles" associated with the products or processes under scrutiny. To take the example of energy systems, life cycle analysis not only would examine the effects due to emissions from particular types of power stations, but also would include attention to the emissions associated with the transportation of fossil fuels, heavy metal discharges associated with oil extraction or mining, and the pollution associated with the embodied energy associated with materials like the steel and concrete from which wind turbines are constructed. Life cycle analysis typically also considers the impacts incurred during the construction phase of the facilities in question, as well as those due to the eventual decommissioning of the plant and the management of any residual arisings. Accordingly, the system boundary protocols developed in life cycle analysis can be applied in conjunction with a variety of other appraisal methods, including probabilistic risk assessment or benefit–cost analysis.

4.3.10 SOCIAL IMPACT ASSESSMENT

Social impact assessment (SIA) is typically defined as "efforts to assess or estimate, in advance, the social consequences that are likely to follow from specific policy actions (including programs, and the adoption of new polices), and specific government actions" (Interorganizational Committee 1994:2). SIA is somewhat parallel to environmental impact assessment although it focuses on social and cultural impacts and examines these impacts using a variety of social science methods. The components of SIA include 1) involving all affected members of the public; 2) identifying winners and losers and particularly vulnerable groups, examining public concerns; 3) defining significance in terms of impacts and outlining assumptions; 4) providing feedback to planners and managers; 5) establishing monitoring programs; and 6) employing trained social scientists to identify and assess the impacts using the best available data sources.

SIA can overlap with some elements of economic impact assessment, especially in terms of employment impacts. However, social impact assessment routinely involves quantitative as well as qualitative research methods in examining impacts. See Turnley et al. in this volume for a more detailed discussion of SIA.

4.4 EVALUATION CRITERIA

In this section we propose a set of evaluation criteria to assess the various methods described in section 4.3. Since the methods are very broad, we also choose a very broad set of evaluation criteria. We do not focus on detailed methodological issues associated with the techniques. For example, we do not evaluate methods in terms of statistical performance. We also do not provide a detailed evaluation of methods within a category. Stated preference methods, for example, are evaluated as a whole to avoid delving into the detail of random sample contingent valuation approaches versus deliberative approaches resulting in rankings or alternatives. However, we do provide an assessment of whether the methods have potential to perform well on a given criterion or whether the method does not have such potential.

4.4.1 THE ABILITY TO CAPTURE HETEROGENEITY

A method is deemed to be superior if it can capture and represent the heterogeneity in preferences or values over space and time. Heterogeneity over space can be thought of as differences in preferences across people in different geographic locations or people in different income and demographic classes. Being able to identify such heterogeneity is important to provide decision makers with the distributional implications of their decisions. Such information may help decision makers consider remedial or mitigation actions as a result of inequitable impacts of a policy decision. Methods that characterize heterogeneity also tend to be more robust in developing measures of value. For example, if preferences are highly skewed (a good is strongly preferred by half of the sample but disliked by the other half), a statistical approach that ignores heterogeneity will generate the result that the good is uniformly neutral to the sample.

Capturing and representing heterogeneity over time are more challenging. Ideally, an approach should be able to identify preference evolution either at an individual level (e.g., preferences changing in response to age) or at the population level (e.g., aggregate demand changing with an aging population). Preference evolution as a result of learning, however, is difficult to capture. Approaches that help reflect the implications of today's actions and policies on future environmental, social, and economic conditions are most desirable.

The degree to which a method captures heterogeneity affects the level at which the results can be aggregated. Aggregation — the process of combining information across different dimensions temporally (time frame), spatially (geographic frame), or socially (e.g., demography) — is more defensible with methods that capture heterogeneity to a higher degree. The level of aggregation possible varies within a given method and across different types of methods. The ability of a method to allow aggregation is partially a function of the way results are reported for a given method used, and partially a function of the representativeness of information used in the method. This creates an integral link between aggregation and sampling issues. For example, the ability to aggregate the results of a method that monetizes environmental values across different socioeconomic strata is dependent on the representativeness of the data used and the specific techniques of analysis used to adequately characterize the populations of concern.

Sampling is the use of subsets of information to represent a larger population of concern. Sampling can occur across individuals, environmental services, or physical or temporal locations. While all of the methods evaluated use some form of sampling, the influence of that sampling varies across approaches. A fundamental issue of sampling is the degree to which the chosen sample adequately represents the population or information of concern. Biased (e.g., unrepresentative) sampling limits the ability to adequately capture heterogeneity and thus aggregate information at higher levels.

Table 4.1 contains an evaluation of the various methods and their ability to capture heterogeneity and to aggregate. Note that we refer to heterogeneity among the observers or respondents or citizens and not to heterogeneity in the approaches or natural systems being examined. Stated preference methods score medium and high

TABLE 4.1
An assessment of valuation methods

Valuation methods	Heterogeneity (observer)	Complexity (observed)	Uncertainty (risk [R], uncertainty [U], ignorance, ambiguity [A], or deterministic [D])	Ordinal (O) and cardinal (C) outputs	Responsiveness to context, human and natural (manipulable)	Practicality (applicability) to risk assessment tiers and decision context), screening and definitive
Economic methods						
Stated preference methods[a]						
CV/SP	M*[b,c]	M*	R, U	C	M*/M*	No/Yes
Multicriteria analysis	H	M*	R, U, A	O	M*/M*	No/Yes
Revealed preference methods[a]	M*	L*	R	C	L*/L*	No/Yes
Green gross domestic product (GDP)	L	M	R	C	L*/L*	No/Yes[d]
Economic impact analysis	L	L*	R	C	L*/L*	No/Yes[e]
Replacement cost and opportunity cost	L	L	R	C	L*/L*	No/Yes
Habitat equivalency analysis	L (at best)	L*	R	C	L*/M	Yes/Yes
Energy-based methods	Absent	M*	D*	C	L*/M	Yes/Yes

(*continued*)

TABLE 4.1 (CONTINUED)
An assessment of valuation methods

Valuation methods	Heterogeneity (observer)	Complexity (observed)	Uncertainty (risk [R], uncertainty [U], ignorance, ambiguity [A], or deterministic [D])	Ordinal (O) and cardinal (C) outputs	Responsiveness to context, human and natural (manipulable)	Practicality (applicability to risk assessment tiers and decision context), screening and definitive
Indices						
Ecological sustainability index	L	M	D*	C	L*/L*	No / Yes[e]
Ecological footprint	L	L	D*	C	L*/L*	No / Yes
Genuine progress indicator (GPI)	L	M	D*	C	L*/L*	No / Yes[e]
Mass balance and natural step	L	M	D*	C	L/L?	?
Life cycle assessment	L*	M*	R	O	L*/M	No / Yes
Technology assessment	M*	M*	R, U	O	L*/M	No / Yes
Social impact assessment	H?	M?	R, U?	O	L*/M	?

[a] Stated preference methods are more susceptible to manipulation or strategic behavior than are revealed preference methods. Both are sensitive to context that may result in manipulation. Both of these forms of manipulation are undesirable.
[b] Letter codes followed by an asterisk (*) indicate that the method can perform better than this rating but the typical application performs at the level indicated in the table.
[c] L = low degree. M = medium degree. H = high degree.
[d] National-level analysis only.
[e] Regional-level analysis.

Valuation Methods

on this scale as they are typically designed to identify preferences from a variety of types of individuals and the methods can present disaggregate preference information. Furthermore, since these methods are based on economic measures and permit monetization, they can be aggregated within an economic framework (sum of consumer surplus measures). Multicriteria analysis methods even more discriminating on this measure as individuals can frame the situation in their own ways, permitting even more freedom to identify heterogeneity. A potential downside of decision analysis methods is that they are not typically conducted with large representative samples, reducing their ability to reflect a complete picture of heterogeneity. Revealed preference methods are also able to capture heterogeneity and can aggregate in a consistent fashion. Economic measures already based on aggregation (such as EIA) have more difficulty presenting information on heterogeneity. Thus, they are scored lower in our assessment. Metrics based primarily on natural systems (HEA and energy analysis) do not reflect information on preferences to any great degree; thus, they do not perform well in terms of heterogeneity. Similarly, highly aggregated indices (ESI and EF) contain a form of preference information, but they do not generate an understanding of heterogeneity. These indexes are often based on measures that arise from strong sampling procedures, but they are highly aggregated. The aggregation across categories is often based on rather arbitrary weightings, relative to economic metrics, making these methods inconsistent with benefit–cost frameworks. However, the weighting approach allows for the construction of a single metric that includes much more than economic methods.

4.4.2 The Ability to Incorporate Complexity of Natural Systems

Ecological risk assessments often involve examination of complex systems (human and ecological), and thus the valuation of changes in these systems should be sensitive to their complexity. Ecological service flows, for example, are often characterized by complex interactions and potential thresholds. Understanding the implications of changes to these service flows will involve the understanding of these interactions. Valuation techniques vary in the extent to which they can capture such complexities. In Table 4.1, we provide an assessment of how well each valuation technique can capture the complexity of the observed system and provide value measures that are sensitive to this level of complexity.

The ratings provided in Table 4.1 illustrate that several techniques have the ability to handle complexity. Some of the traditional economic approaches, however, are somewhat limited in this regard. Revealed preference approaches, for example, do not typically include models of ecosystem function and thus are limited in their ability to capture the complexities of ecosystems. Replacement cost methods abstract from ecosystem complexity and examine only the cost of the replacement alternatives. The ecological footprint also scores poorly on this scale since it converts ecological and human system components into a single land use metric. Overall, however, many of the techniques perform well because they are based on approaches that can provide insight into complex systems (e.g., stated preference approaches), or are based on models or representations that capture a significant degree of complexity (e.g., energy-based methods).

4.4.3 THE HANDLING OF INCOMPLETE KNOWLEDGE IN DECISION MAKING

This chapter has discussed the wide variety of different sources of risk and uncertainty associated with the characterization of complex intercoupled ecological and economic systems. There exists a formidable array of appraisal techniques and tools for addressing such uncertainties and rendering them more tractable and meaningful to decision makers. However, although highly diverse in form, provenance, and context, the implications of these uncertainties for decision making can be understood in terms of 4 discrete "states of knowledge" — each associated with a distinct set of assessment and decision-making responses. The first is the condition of *risk*, under which it is possible fully to characterize a range of possible outcomes and to determine the relative likelihood of each. The second is the condition of uncertainty (in the strict sense of this term), under which the range of possible outcomes remains fully characterized but there exists no firm basis for confidence concerning their relative likelihoods. A third condition is ambiguity, under which the challenge concerns not the likelihood but the characterization, partitioning, or bounding of the possible outcomes themselves. The 4th and final condition is referred to as ignorance, where not only the likelihoods but also some of the possibilities themselves may be unknown — where we face the prospect of surprise. The interrelationships between the definitions of these 4 terms are represented in Figure 4.1. Together, they constitute a complete and coherent framework for understanding the states of knowledge associated with different degrees of incertitude. The practical responses associated with each state of knowledge are discussed below in a little more detail.

The condition of risk exists where the systems in question are relatively well understood, where there exist a long base of relevant experience and an expectation that possible future circumstances will resemble past conditions, or where there is confidence in the completeness, robustness, and specificity of theoretical models (Keynes 1921; Knight 1921). It is under these conditions that incertitude can be well characterized for decision makers through the use of probabilistic methods (Luce and Raiffa 1957; Morgan et al. 1990). Where the different possible states of the world are understood in terms of distinct categories, then their relative likelihoods can be expressed as discrete scalar probabilities (Burgman 2005). Where outcomes are ordered according to continuous performance parameters, then probability density functions can be used (Evans 1986; Bailar 1988). Either way, a range of Bayesian procedures provides a means to update prior information or beliefs in the light of new research or monitoring information (Jaynes 1986; Wallsten 1990). Delphi techniques allow the aggregation of a variety of expert judgments (Martin and Irvine 1989). Based on these kinds of data, the disciplines of probabilistic and comparative risk assessment have developed a host of more specific methods for the analysis of different dose–response relationships and the ranking and comparison of different types of probability distribution (Goodman 1986; Beck 1987).

Under the strict definition of the state of uncertainty (uncertainty in this context is that which we do know, or which we are unsure of), however, these kinds of sophisticated probabilistic techniques are — by definition — not applicable. Although the term "uncertainty" is used in different ways in various branches of risk assessment and economics, this remains the original and most authoritative definition.

Valuation Methods

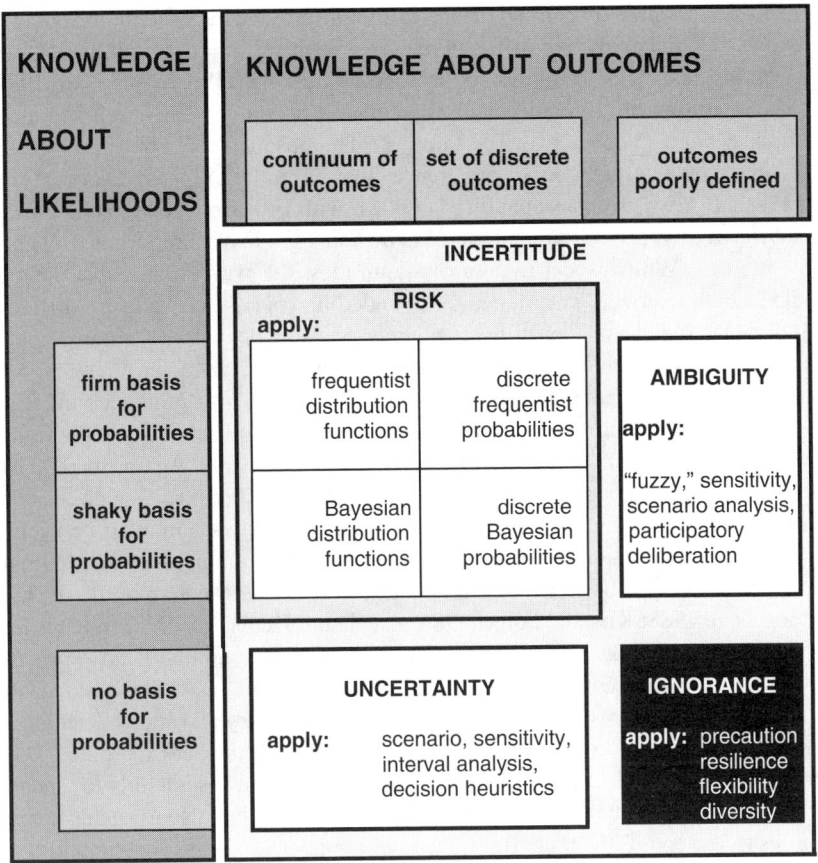

FIGURE 4.1 Contrasting degrees of incomplete knowledge (Stirling 1999).

Whatever it is called, Figure 4.1 shows how this condition of uncertainty is (along with the other states of knowledge) fundamentally implied by the formal definition of risk itself. Attachment to the elegance and facility of probabilistic analysis sometimes prompts a desire to deny this sense of the term "uncertainty" (Lumby 1984; McKenna 1986; Morgan et al. 1990). However, such aspirations have long been refuted in probability theory, where leading practitioners such as de Finetti (1974) emphasize that "probability does not exist" under such conditions. Likewise, leading economists are also well aware of the limits of probabilistic methods, such as Friedrich Hayek, who referred in his Nobel Prize acceptance speech to this tendency to the denial of uncertainty as the "pretence at knowledge" (1978). Fortunately, though probabilistic techniques remain inapplicable, there are a variety of robust methodological responses to uncertainty. In particular, sensitivity (Amendola et al. 1992; Saltelli 2001), scenario (Berkhout et al. 2001), and interval (Moore 1966) analysis can be used to explore the implications of different possibilities, without knowing their relative likelihood. The numeral unit spread assessment pedigree (NUSAP) procedure proposed by Funtowicz and Ravetz (1990) offers a further framework for analysis. A range of heuristic rules of thumb provides a systematic framework for expressing and analyzing different reactions to uncertainty, including the "maximin," "minimax," and "minimax regret" decision criteria (Pearce and Nash 1981).

The condition of ambiguity applies where we face a challenge, not so much of uncertainty but of "contradictory certainties" (Thompson and Warburton 1985). This arises from the heterogeneity of preferences that have already been discussed. Even past events (for which the probability is known with full confidence to be 1) may be the subject of divergent interpretations as to meanings, salience, importance, or appropriate metrics. With respect to future possibilities, the way the possible outcomes should be characterized, partitioned, or bounded is typically viewed very differently by individual specialists, scientific disciplines, stakeholder interest groups, political constituencies, and sociocultural communities. The open character of future events compounds the divergent understandings and raises the stakes (Funtowicz and Ravetz 1990). Even within a particular body of analysis, quite subtle changes in the framing assumptions can often yield quite massively divergent results. Again, there exists a wide variety of analytical and procedural responses to ambiguity. As with the condition of uncertainty, scenario, sensitivity, and interval analyses may all be useful in exploring the implications of divergent framings. Where ambiguities are dominated by dual polarized positions, then the techniques of fuzzy set theory may also be useful (Dubois et al. 1988; Klir and Folger 1988; Zadeh and Kacprzyk 1992). What is likely to be more crucial, however, are the institutional settings and deliberative processes under which the different perspectives are elicited in appraisal. As discussed in Chapter 2, this is where inclusive approaches for transdisciplinary and stakeholder engagement and citizen and community participation come to the fore (Joss and Durant 1995; Renn et al. 1995; Burgess et al. 2001). These offer means not only to render the decision-making process more democratically legitimate, but also to enable the more rigorous treatment of divergent framing assumptions in analysis (Stirling 2003).

The 4th and final state of knowledge encountered in decision making on environmental risk is the state of ignorance. This is a condition under which we are able *neither* fully to quantify likelihoods *nor* definitively to characterize, partition, or bound all the possible outcome parameters (Ravetz 1987; Wynne 1992; Faber et al. 1994; Stirling 1998). Although frequently neglected in risk assessment, the state of ignorance is widely acknowledged elsewhere in economics (Keynes 1921; Shackle 1968; Loasby 1976; Ford 1983; Dosi and Egidi 1987), decision theory (von Winterfeldt and Edwards 1986; Winkler 1986), technology assessment (Collingridge 1980, 1982), and environmental policy (Smithson 1989). Put at its simplest, ignorance is a reflection of the degree to which "we don't know what we don't know." It represents our uncertainty about our uncertainty (Cyranski 1986). It is an acknowledgment of the importance of the element of surprise (Brooks 1986).

Crucially, ignorance does not necessarily imply the complete absence of knowledge, but simply the possibility that certain relevant parameters may be unknown or unknowable. This was the case at various times, for instance, in the history of recognition of endocrine disruption mechanisms (Thornton 2000) and bovine spongiform encephalopathy (BSE; van Zwanenberg and Millstone 2001). Ignorance may operate at a more localized institutional (rather than societal) level, when pertinent knowledge is in existence, but remains unused in decision making. This was the case, for instance,

Valuation Methods

in the early stage of experience with occupational and public health hazards such as asbestos (Gee and Greenberg 2001), benzene (Infante 2001), and various organochlorine chemicals (Ibaretta and Swan 2001), as well as with stratospheric ozone depletion (Farman 2001).

Although somewhat marginalized by a preoccupation with the probabilistic methods of risk assessment, there remains a range of robust practical responses to ignorance. The precautionary principle, for instance, is strongly contested as a firm decision rule, but like other recent approaches within risk assessment suggests a range of concrete policy responses (NRC 1996; USPCS 1997; RCEP 1998; GMSRP 2003). These include considering a wide range of effects, involving a broad variety of disciplines and interests, comparing the pros and cons of different options and potential substitutes, exploring a full array of all scenarios and possibilities, prioritizing research and monitoring, and attending explicitly to the setting of levels of proof and burdens of persuasion in regulatory appraisal (Stirling 1999; Gee et al. 2001; Renn et al. 2003). More specifically, there exist a range of particular strategic responses to ignorance such as the diversification of portfolios (Stirling 1994, 1998), the pursuit of flexibility in technological trajectories (Collingridge 1983), and the avoidance of persistent, ubiquitous, or irreversible effects (Mueller-Herold et al. 2003). These offer a variety of concrete ways to be more humble about the role of science in the management of risk (Raffensberger and Tickner 1999; O'Brien 2000). In essence, this is simply an expression of the sound scientific principle that "knowing one's ignorance is the best part of knowledge" (Morgan et al. 1990). Science remains an absolutely essential input, but it is a necessary, rather than a sufficient, condition for effective decision making under ignorance (Stirling 2003).

Table 4.1 contains an evaluation of the various valuation methods in terms of their ability to incorporate the 4 dimensions of risk, uncertainty, ambiguity, and ignorance. If a method is able to include 1 or more of these dimensions into its approach, the corresponding letter is placed in the row. If the method is inherently deterministic, and by definition unable to include uncertainty, the letter "D" is placed in the row. Methods that are based on preference elicitation can capture risk, uncertainty, and in some cases ambiguity. Most economic valuation methods can capture uncertainty. Many biologically oriented methods, however, do not typically include risk considerations. These methods could include risk considerations; thus, the D rating is modified to "D*" to indicate that the method could potentially go beyond this level.

4.4.4 ORDINAL VERSUS CARDINAL MEASURES

Ordinal measures are used to rank outcomes, and only the ranking of the outcomes matters. The distances between the measures are not meaningful. More or less, higher or lower, and sooner or later are ordinal measures. There may be numbers associated with the measure, but the differences between these numbers are not relevant. For example, a scaling exercise in which something is ranked as high, medium, or low might assign numbers to these (high = 1, medium = 2, and low = 3), but might as well have used any other numbers that maintain the same ordering. In cardinal measures, the differences on the measure matter. For example, height measured as "taller" is ordinal, but measured in inches is cardinal. This distinction is important, based on what mathematical operations are legitimate on the resulting scale, as well as the information content they contain and the difficulty (cost or controversy) of obtaining the measure. It is not legitimate to add ordinal measures, and yet this occurs all the time in indices that rank several criteria by ordinal scales, and then total or average the scores.

Individual preferences are often held to be ordinal by economists. Thus, preferences are ordinal rankings of which collection of goods, including environmental goods, is judged to be preferred. We cannot add ordinal preferences, and hence cannot just add together individual rankings to get to a social or group ranking. More information is needed. Willingness to pay, or monetization, is a particular "cardinalization" of the preferences so that they can be added. If, instead of asking which bundle is preferred, we ask how much of a particular good or money one would give up to have one bundle over another, we have added information that allows addition. Of course, this is but one possible legitimate cardinalization of preferences, and one that may or may not have special sway. It is the one that is used in benefit–cost analysis, but this does not have unique legitimacy.

One evaluative criterion for the index or approach is whether it provides an ordinal or cardinal representation of the underlying ranking. More can be done with cardinal indices, but more must be given up to attain them — yet another instance of the "no free lunch" axiom of economics!

Whether the method results in an ordinal or cardinal measure is depicted in Table 4.1. All the methods result in cardinal measures of effects, except for multicriteria analysis, life cycle, technology, and social impact assessments. Those that result in ordinal measures can be used only to rank outcomes across alternatives.

4.4.5 Responsiveness to Context

Some valuation methods are highly dependent on the context of the measurement and the framework used to implement the method, while others are relatively unresponsive to changes in the decision context. We evaluate the methods based on 2 aspects of context responsiveness: human and natural. "Human context" refers to the choices made by those implementing the process and includes the array of assumptions and framing choices made by the investigator. For example, in the contingent valuation method, this would include the method of payment for the good. "Natural context" responsiveness refers to the extent that the method is responsive to changes in the social-ecological system, and is a measure of the inclusiveness or scope of the technique.

The economic methods tend to be highly context dependent, on both human and natural dimensions. There is not a single, enduring economic value for environmental goods, as an individual's preferences may change over time. Moreover, since preferences differ across people (heterogeneity), the social or group value will depend on the aggregation method (i.e., on the decision process).

But context dependence of economic measures goes beyond this. The amount people are willing to pay depends on what they have to pay and when, how, and a rule for who else pays. As technology changes, their information about the good changes, all other manner of influences change, and the value expressed changes. Much of this sensitivity is expected and shows that the method is working well; failure to respond to context changes can be used to show a problem with the method or its implementation (Mitchell and Carson 1989).

Within the economic methods, SP (and experiment-based RP) techniques are most highly context dependent. This has the advantage of allowing them to be applied to states of the world that have not existed, since the analyst can create hypothetical worlds (within limits of the respondent's ability or desire to accept counterfactuals). The RP methods, since they use data on behavioral responses to existing and/or past conditions, are less context dependent, but there is still considerable scope for context response (e.g., what sites are alternative destinations, and assumptions about the cost of time in a travel cost model). Noneconomic valuation methods tend to be less responsive to human contextual issues. For example, emergy, the ecological footprint, and the various indices will respond to changes in the system, but not to changes in how people feel about the system. The indices (e.g., the ecological footprint) respond to a large array of elements within their purview, but that purview is limited (e.g., not including the social component).

Whether the method is context sensitive is different than the degree to which these sensitivities are highlighted in implementation. The open-closed evaluative criterion (see above) indicates the extent to which sensitivity to frame or context or another assumption is explored fully in the implementation of the method and presented to decision makers, or is hidden or assumed away via choices of the analyst.

The ability of a measurement method to respond to context generally is a desirable feature of the method, but can be abused. Thus, those who can respond to context also are prone to strategic manipulation. This can be done by the analyst, if he or she understands that certain choices (of frames or assumptions) lead to certain desired (by the analyst) outcomes. This, of course, is true to an extent of any method embodying assumptions (as they all do). But, when very subtle frame and context specifications lead to quite different results, this expands the potential for misapplication or misrepresentation.

The stated preference methods are uniquely vulnerable to manipulation by the subjects being observed by the analysts. By recognizing the relationship between answers given and ultimate outcomes, a subject can try to manipulate the process to his or her advantage. This is known as strategic bias. Those methods that rely on secondary data are more strategy proof.

Methods like "life cycle analysis" and technology assessment as typically implemented are limited in responsiveness to human context factors, but are more responsive to natural context changes. There exists potential for these methods, and the index methods, to be more sensitive to context, but this is not the way they are commonly implemented.

4.4.6 PRACTICALITY

In order to evaluate the relevant resource valuation techniques and methods, a "practicality" standard was applied. Practicality judgments focused on a specific method's applicability within a conventional risk assessment framework and within a variety of anticipated decision-making contexts. The practical application and implementability of each method at various scales and specific "screening-level" versus "definitive" tiers were assessed and are presented in Table 4.1. Elements of the practicality analysis included methodological complexity, information and data requirements, ease of communication and presentation, utility, relevance, and the degree to which the specific method is practiced in the field.

With the exceptions of the Habitat Equivalency Analysis and emergy methods, none of the valuation methods was considered practical for implementation for the majority of screening-level risk assessment purposes. Most of these methods require primary data collection, to one degree or another, and any significant data collection efforts are contrary to the intent of most screening-level assessments. Habitat Equivalency Analysis and emergy methods, however, provide elements of resource valuation even at the screening-level stage and therefore may be practical to implement.

Depending upon the specific decision support requirements, any of the valuation methodologies may be practical to implement for definitive risk assessment purposes, as indicated in Table 4.1. The primary data collection requirements and costs for implementing the methods may be justified by the scale of the definitive risk assessment or the criticality of the anticipated risk management decisions. While most of the valuation methods can be applied at nearly any risk assessment scale, the green GDP method can only be applied at the regional-national level consistent with its intended objective. Economic impact analysis, the Ecological Sustainability Index, and the Genuine Progress Indicator may each be practically applied in support of definitive risk assessments at the regional level of analysis.

4.5 PLUG AND PLAY TOOLS: EVALUATION AIDS THAT CAN BE APPLIED TO ALL METHODS

A number of evaluation aids or analysis tools can be applied to all of the methods described in the chapter. Often these are only included in certain types of analysis, but they generalize to almost all methods. This section describes a few of these tools, recognizing that a more detailed discussion of these issues has been carried out in other chapters. Discounting is the first of these issues. Discounting can be applied to money or physical attributes to characterize weights for future versus present outcomes. Discounting can be subjected to a sensitivity analysis to determine how sensitive outcomes are to the weighting applied to present versus future levels of attributes. Sensitivity analysis is also a tool that can be applied to all of the methods described in this chapter. Sensitivity analysis around plausible parameter values provides a characterization of the response surface and can provide significant information about critical factors and areas where further research could provide significant information value. A variety of mathematical and statistical methods can also be applied to enhance the methods. These include Bayesian approaches to processing uncertainty, fuzzy logic methods, and others. A description of some of these methods is provided in Appendix 4.2.

4.6 CONCLUSIONS

In the final assessment, several economic and noneconomic methods are available to support the valuation of ecological resources. Of these, preference methods and decision analysis are suggested by our evaluation to have the greatest potential to perform well in a broad range of situations and decision contexts (to the degree that our familiarity with the potential performance of these methods and comfort with a benefit–cost framework for analysis haven't biased the assessment). However, other methods might be better suited to particular problems, decision frames, and the values of the decision maker. Yet, with the exceptions of the Habitat Equivalency Analysis and emergy evaluation methodologies, we conclude that the methods were generally impractical to implement at the screening level of risk assessment. Alternatively, depending upon the scale of analysis and criticality of the risk management decisions, any of the valuation methods could be practically implemented at the definitive risk assessment level. A criterion not evaluated in Table 4.1 is one of capacity, the availability of skilled practitioners to perform the method. As the demand for integrated economic and ecological assessment grows, we expect a concomitant increase in the supply of those practitioners.

We also expect that methodological improvements will continue, in terms of both resolving the limitations of current approaches and developing new approaches and methods. The asterisks in Table 4.1 reflect opportunities for sophisticated applications of existing methods, but these typically are not made. Certain decisions may require enhanced levels of sophistication, providing the pressure for refinement and improvement of those methods. A particular weakness in all of the valuation methods is that they are based on static frameworks, while the social and ecological systems are inherently dynamic. Future research on valuation in dynamic systems is critically required.

In conducting the evaluation of methods, it became evident that economists and ecologists utilize different approaches for problem solving (Figure 4.2). Economists focus their analysis on the human system. Such an approach is prone to error arising from a lack of understanding and integration of the natural system. Conversely, ecologists focus on the natural system in the absence of people's values, which increases the susceptibility to error about those values when they are relevant to the situation. We conclude that a more robust alternative to both these approaches is an integrated one in which people's values about nature are the analytical frame. By this means, error about both the natural system and people's value should be minimized. We recommend integrated approaches to reduce errors in the valuation process arising from misunderstandings about natural systems and human systems.

Reflecting the inherent complexity of ecological and social systems, decision makers should recognize that many problems are best informed by the results of open analytic approaches — for these situations, there are no simple answers. In contrast to closed approaches, which focus the analysis on single numbers or outcomes, open approaches allow for and present to decision makers the plurality of values and assumptions, opportunities for learning, and so on. Risk assessment, benefit–cost analysis, and other approaches to health, environmental, and wider impact assessment are conventionally conducted as a means to provide concrete prescriptive policy recommendations. It is well understood that such methods can only offer aids to,

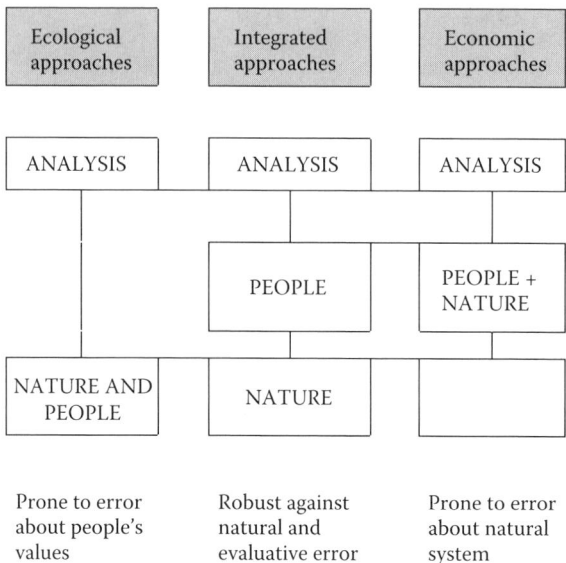

FIGURE 4.2 Analysis approaches used by economists and ecologists.

rather than substitutes for, decision making. Constraining these approaches to provide "single-number answers" rather than employing them in open processes, however, will severely limit the degree to which these methods can aid decision making. Open approaches to analysis will significantly enhance our ability to understand the consequences of alternative decisions.

REFERENCES

Adamowicz WL, Louviere JJ, Swait JD. 1998. An introduction to stated choice methods for resource based compensation. Prepared by Advanis Inc. for the National Oceanic and Atmospheric Administration, US Department of Commerce. Edmonton (AB): Advanis.

Amendola A, Contini S, Ziomas I. 1992. Uncertainties in chemical risk assessment: results of a European benchmark exercise. J Haz Mat 29:347–363.

Bailar J. 1988. Scientific inferences and environmental problems: the uses of the statistical thinking. Chapel Hill: Institute for Environmental Studies, University of North Carolina.

Beck M. 1987. Water quality modeling: a review of the analysis of uncertainty. Water Resour Res 23:1393–1442.

Berkhout F, Hertin J, Jordan A. 2001. Socio-economic futures in climate change impact assessment: using scenarios as "learning machines." Tyndall Centre Working Paper No. 3. Norwich, UK: Tyndall Centre for Climate Change Research.

Boyce R, Brown TC, McClelland GH, Peterson GL, Schultze WD. 1992. An experimental examination of intrinsic values as a source for the WTA-WTP disparity. Amer Econ Rev 82:1366–1373.

Boyle K, Morrison M, Taylor L. 2004. Why value estimates generated using choice modeling exceed contingent valuation: further experimental evidence. Paper presented at the Australian Agricultural and Resource Economics Society Conference, Melbourne, February.

Boyle, KJ. 2003. Contingent valuation in practice. In: Champ P, Boyle KJ, Brown TC, editors, A primer on nonmarket valuation. Dordrecht (the Netherlands): Kluwer, p 111–170.

Brooks H. 1986. The typology of surprises in technology, institutions and development. In: Clark WC, Munn RE, editors, Sustainable development of the biosphere. Cambridge (UK): Cambridge University Press for the International Institute for Applied Systems Analysis, p 325–348.

Brown TC, Peterson G. 2003. Multiple good valuation. In: Champ P, Boyle KJ, Brown TC, editors, A primer on nonmarket valuation. Dordrecht (the Netherlands): Kluwer p 221–258.

Burgess J, Clark J, Stirling A, Studd K, Chilvers J, Lewis S. 2001. Local outreach. R&D Technical Report SWCON 204. Bristol (UK): Environmental Agency.

Burgman M. 2005. Risks and decisions for conservation and environmental management. Cambridge (UK): Cambridge University Press.

Carson RT. 2000. Contingent valuation: a user's guide. Env Sci Technol 34:1413–1418.

Chapman D, Iadanza N, Penn T. 1998. Calculating resource compensation: an application of the service-to-service approach to the Blackbird Mine hazardous waste site. NOAA Damage Assessment and Restoration Program, Technical Paper 97-1. Washington (DC): National Oceanic and Atmospheric Administration.

Checkland P. 1981. Systems thinking, systems practice. London (UK): John Wiley.

Chuenpagdee R, Knetsch JL, Brown TC. 2001. Environmental damage schedules: community judgments of importance and assessments of losses. Land Econ 77:1–11.

Cobb C, Glickman M, Cheslog C. 2001. The Genuine Progress Indicator 2000 update. Redefining Progress. December. http://www.rprogress.org/projects/gpi/. Accessed October 5, 2003.

Collingridge D. 1980. The social control of technology. London (UK): Francis Pinter Press. 200 p.

Collingridge D. 1982. Critical decision making: a new theory of social choice. London (UK): Frances Pinter. 189 p.

Collingridge D. 1983. Technology in the policy process: controlling nuclear power. London (UK): Frances Pinter. 254 p.

Conrad J, Clark C. 1987. Natural resource economics: notes and problems. Cambridge (UK and NY): Cambridge University Press. 231 p.

Costanza R. 2000. Commentary forum — the ecological footprint: the dynamics of the ecological footprint concept. Ecol Econ 32:341–345.

Cyranski J. 1986. The probability of a probability. In: Justice JH, editor. Maximum entropy and Bayesian methods in applied statistics. Cambridge (UK): Cambridge University Press, p 106–116.

Davies G, Burgess J, Eames M, Mayer S, Staley K, Stirling A, Williamson S. 2003. Deliberative mapping: appraising options for addressing "the kidney gap." Final report to Wellcome Trust. http://sss.deliberative-mapping.org.

de Finetti N. 1974. Theory of probability. New York (NY): Wiley.

Dosi G, Egidi M. 1987. Substantive and procedural uncertainty: an exploration of economic behaviours in complex and changing environments. SPRU DRC Discussion Paper No. 46. Sussex, UK: Science and Technology Policy Research, University of Sussex.

Dubois D, Prade H, Farreny H, Martin-Clouaire R, Testemale C. 1988. Possibility theory: an approach to computerized processing of uncertainty. New York (NY): Plenum. 263 p.

Ecological Economics. 2000. Special issue. 32(3).

Evans N. 1986. Assessing the risks of nuclear energy. In: G. Harrison, D. Gretton, editors, Energy UK policy journals. Newbury (UK): Energy UK.

Faber M, Proops J, Manstetten R. 1994. Evolution, time, production and the environment. Berlin: Springer-Verlag. 288 p.

Farman J. 2001. Halocarbons, the ozone layer and the precautionary principle. In: Gee D, Harremoes P, Keys J, MacGarvin M, Stirling A, Vaz S, Wynne B, editors, Late lesson from early warnings: the precautionary principle 1898–2000. Copenhagen

(Denmark): European Environment Agency, p 76–83. http://reports.eea.europa.eu/environmental_issue_report_2001_22/en/issue-22-part-07.pdf. Accessed May 7, 2007.

[FFF] Forum for the future, 2003. The mass balance UK project. http://www.massbalance.org. Accessed October 5, 2003.

Ford J. 1983. Choice, expectation and uncertainty: an appraisal of G.L.S. Shackle's theory. Totowa (NJ): Rowman & Littlefield.

Freeman AM. 2003. The measurement of environmental and resource values. Baltimore (MD): Resources for the Future. 489 p.

Funtowicz S, Ravetz J. 1990. Uncertainty and quality in science for policy. Dordrecht (the Netherlands): Kluwer. 229 p.

Gee D, Greenberg M. 2001. Asbestos: from "magic" to malevolent mineral. In: Gee D, Harremoes P, Keys J, MacGarvin M, Stirling A, Vaz S, Wynne B, editors, Late lesson from early warnings: the precautionary principle 1898–2000, European Environment Agency, Copenhagen, p 52–63. http://reports.eea.europa.eu/environmental_issue_report_2001_22/en/issue-22-part-05.pdf. Accessed May 7, 2007.

Gee D, Harremoes P, Keys J, MacGarvin M, Stirling A, Vaz S, Wynne B. 2001. Late lesson from early warnings: the precautionary principle 1898–2000. Copenhagen (Denmark): European Environment Agency.

Goodman J. 1986. On criteria of insignificant difference between two risks. Risk Anal 6:235–244.

Grin J, van de Graaf H, Hoppe R. 1997. Technology assessment through interaction: a guide. The Hague (the Netherlands): Rathenau Institute. 98 p.

Hayek F. 1978. New studies in philosophy, politics, economics and the history of ideas. Chicago: Chicago University Press. 314 p.

Holmes T, Adamowicz WL. 2003. Attribute based methods. In: Champ P, Boyle KJ, Brown TC, editors, A primer on nonmarket valuation. Dordrecht (the Netherlands): Kluwer, p 171–220.

Ibaretta D, Swan S. 2001. The DES story: long term consequences of prenatal exposure. In: Gee D, Harremoes P, Keys J, MacGarvin M, Stirling A, Vaz S, Wynne B, editors, Late lessons from early warnings: the precautionary principle 1898–2000. Copenhagen (Denmark): European Environment Agency.

Infante P. 2001. Beneene: an historical perspective on the American and European occupational setting. In: Gee D, Harremoes P, Keys J, Mac Garvin M, Stirling A, Vaz S, Wayne B, editors. Late lessons from early warnings: the precautionary principle 1898–2000. Copenhagen (Denmark): European Environment Agency, p 38–51.

Interorganizational Committee on Guidelines and Principles for Social Impact Assessment. 1994. Guidelines and principles for social impact assessment. Washington (DC): US Department of Commerce, National Oceanic and Atmospheric Administration, National Marine Fisheries Service. 26 p.

Jansson AM, Hammer M, Golke CF, Costanza R. 1994. Investing in natural capital: the ecological economics approach to sustainability. International Society for Ecological Economics. Covelo (CA): Island Press. 504 p.

Jaynes E. 1986. Bayesian methods: general background. In: Justice JH, editor. Maximum entropy methods in applied statistics. Cambridge (UK): Cambridge University Press. p 1–25.

Joss S, Durant J. 1995. Public participation in science: the role of consensus conferences in Europe. London (UK): Science Museum.

Kay J, Regier H, Boyle M, Francis G. 1999. An ecosystem approach for sustainability: addressing the challenge of complexity. Futures 31:721–742.

Keeney RL, Raiffa H. 1976. Decisions with multiple objectives: preferences and value trade-offs. New York (NY): John Wiley. 569 p.

Keynes J. 1921. A treatise on probability. London (UK): Macmillan. 466 p.

Kinzig AP, Starrett D, Arrow K, Aniyar S, Bolin B, Dasgupta P, Ehrlich P, Folke C, Hanemann M, Heal G, Hoel M, Jansson A, Jansson B, Kautsky N, Levin S, Lubchenco J, Maler KG, Pacala SW, Schneider SH, Siniscalco S, Walker B. 2003. Coping with uncertainty: A call for a new science-policy forum. Ambio 32:330–335.

Klir G, Folger T. 1988. Fuzzy sets, uncertainty and information. Englewood Cliffs (NJ): Prentice Hall. 355 p.

Knight F. 1921. Risk, uncertainty and profit. Boston (MA): Houghton Mifflin. 381 p.

Lancaster K. 1966. A new approach to consumer theory. J Pol Econ 74(2):132–155.

Loasby B. 1976. Choice, complexity and ignorance: an inquiry into economic theory and the practice of decision making. Cambridge (UK): Cambridge University Press.

Lotka AJ. 1922. Contribution to the energetics of evolution. Proc Nat Acad Sci USA 8:147–155.

Loveridge D, editor. 1996. Technology assessment. Special issue. Intl J Technol Mgmt 11(5/6).

Luce RD, Raiffa H. 1957. An axiomatic treatment of utility. In: Luce RD, Raiffa H, editors. Games and decisions: introduction and critical survey. New York (NY): John Wiley. 530 p.

Lumby S. 1984. Investment appraisal. 2nd ed. London (UK): Van Nostrand.

Martin B, Irvine J. 1989. Research foresight: priority-setting in science. London: Pinter. 366 p.

Maxwell T, Costanza R. 1995. Distributed modular spatial ecosystem modeling. Special issue on advanced simulation methodologies. Intl J Comp Simul 5(3):247–262.

McFadden D. 1974. Conditional logit analysis of qualitative choice behavior. In: Zarembka P, editor, Frontiers in econometrics. New York (NY): Academic Press, p 105–142.

McKenna C. 1986. The economics of uncertainty. Brighton (UK): Wheatsheaf-Harvester. 129 p.

Miller RE, Blair PD. 1985. Input-output analysis: foundations and extensions. Englewood Cliffs (NJ): Prentice-Hall. 464 p.

Minnesota Planning Environmental Quality Board. 2000. Smart signals: an assessment of progress indicators. St. Paul (MN): Minnesota Planning Environmental Quality Board. 55 p.

Miranda M, Fackler P. 2002. Applied computational economics and finance. Cambridge (MA): MIT Press. 510 p.

Mitchell RC, Carson RT. 1989. Using surveys to value public goods. Washington (DC): Resources for the Future.

Moore, R. 1966. Interval analysis. Englewood Cliffs (NJ): Prentice-Hall.

Morgan M, Henrion M, Small M. 1990. Uncertainty: a guide to dealing with uncertainty in quantitative risk and policy analysis. Cambridge (UK): Cambridge University Press. 332 p.

Mueller-Herold U, Morosini M, Schucht O, Scheringer M. 2003. Precautionary pre-selection of new organic chemicals: a case study on the application of the precautionary principle in the European Union. In: Renn O, Klinke A, Losert C, Stirling A, van Zwanenberg P, Muller-Herold U, Morosini M, Fisher E, editors, The application of the precautionary principle in the European Union. Final document. Brussels (Belguim): European Commission STRATA Program.

[NRC] National Research Council Committee on Risk Characterization. 1996. Understanding risk: informing decision in a democratic society. Washington (DC): Academy Press.

Natural Step. 2003. http://www.naturalstep.org/learn/peer_reviewed_papers.php. Accessed July 12, 2007.

Nicolette JP, Rockel M, Kealy MJ. 2001. Quantifying ecological changes helps determine best mitigation. Pipe Line Gas Ind. (Sept.):52–57.

O'Brien M. 2000. Making better environmental decisions: an alternative to risk assessment. Cambridge (MA): MIT Press. 286 p.

Odum HT. 1971. Environment, power, and society. New York (NY): John Wiley. 336 p.

Odum HT. 1996. Environmental accounting: emergy and environmental decision making. New York (NY): John Wiley. 370 p.

Odum HT. 2000. Emergy evaluation of an OTEC electrical power system. Energy 25:389–393.

Odum HT, Arding JE. 1989. Emergy analysis of shrimp mariculture in Ecuador. Working paper. Narragansett: Coastal Resources Center, University of Rhode Island. 111 p.

Pearce D, Nash C, 1981. The social appraisal of projects: a text in cost-benefit analysis. London (UK): Meutheun.

Penn T, Tomasi T. 2002. Calculating resource restoration for an oil discharge in Lake Barre, Louisiana, USA. Environ Mgmt 29:691–702.

Phaneuf DJ, Smith VK. 2006. Recreation demand models. In: Mäler K, Vincent J, editors, Handbook of environmental economics. Amsterdam (the Netherlands): Elsevier North Holland, p 671–761.

Raffensberger C, Tickner J. 1999. Protecting public health and the environment: implementing the precautionary principle. Washington (DC): Island Press.

Ravetz J. 1987. Usable knowledge, usable ignorance: incomplete science with policy implications. Knowledge 9:87–116.

[RCEP] Royal Commission on Environmental Pollution. 1998. Setting environmental standards, twenty-first report. London (UK): HMSO.

Rees WE. 1992. Ecological footprints and appropriated carrying capacity: what urban economics leaves out. Environ Urban 4:121–130.

Rees WE. 2000. Eco-footprint analysis: merits and brickbats. Ecol Econ 32:371–374.

Renn O, Webler T, Wiedemann P. 1995. Fairness and competence in citizen participation: evaluating models for environmental discourse. Dordrecht (the Netherlands): Kluwer. 400 p.

Renn O, Klinke A, Losert C, Stirling A, van Zwanenberg P, Muller-Herold U, Morosini M, Fisher E, editors. 2003. The application of the precautionary principle in the European Union. Final document. Brussels (Belguim): European Commission STRATA Program.

Rip A, Misa T, Schot J. 1996. Managing technology in society. London (UK): Pinter. 361 p.

Saltelli A. 2001. Sensitivity analysis for importance assessment. Ispra (Italy): EC Joint Research Centre. http://www.ce.ncsu.edu/risk/pdf/saltelli.pdf.

[SEDAC] Socioeconomic Data and Applications Center. 2004. Environmental Sustainability Index. http://www.ciesin.columbia.edu/indicators/esi. Accessed May 5, 2007.

Shackle G. 1968. Uncertainty in economics and other reflections. Cambridge (UK): Cambridge University Press. 267 p.

Smithson M. 1989. Ignorance and uncertainty: emerging paradigms. New York (NY): Springer. 393 p.

Stirling A. 1994. Diversity and ignorance in electricity supply investment: addressing the solution rather than the problem. Energy Pol 22:195–216.

Stirling A. 1998. Risk at a turning point? J Risk Res 1:97–110.

Stirling A. 1999. On "science" and "precaution" in the management of technological risk. Report to the EC Forward Studies Unit, IPTS, Sevilla, EUR19056 EN. Brussels (Belguim): European Science and Technology Observatory. ftp://ftp.jrc.es/pub/EURdoc/eur19056en.pdf. 61 p.

Stirling A. 2003. Risk, uncertainty and precaution: some instrumental implications form the social sciences. In: Berkhout F, Leach M, Scoones I, editors, Negotiating environmental change: new perspectives from social science. Cheltenham (UK): Edward Elgar. p 33–76.
Taylor LO. 2003. The hedonic method. In: Champ, P., Brown T, Boyle K. editors, A primer on nonmarket valuation. Dordrecht (the Netherlands): Kluwer. p 331–394.
Thompson M, Warburton M. 1985. Decision making under contradictory certainties: how to save the Himalayas when you can't find what's wrong with them. J App Sys Anal 12:3–34.
Thornton J. 2000. Pandora's poison: on chlorine, health and a new environmental strategy. Cambridge (MA): MIT Press. 599 p.
Tschirhart J. 2000. General equilibrium of an ecosystem. J Theor Biol 203:13–32.
Unsworth RE, Bishop RC. 1994. Assessing natural resource damages using environmental annuities. Ecol Econ 11:35–41.
[US PCC] US Presidential/Congressional Commission on risk assessment and risk management. 1997. Framework for environmental health risk management. Final report. Vol. 1. Washington (DC): USEPA.
van den Berg N, Dutilh C, Huppes G. 1995. Beginning LCA: a guide to environmental life cycle assessment. Rotterdam (the Netherlands): CML.
van Zwanenberg P, Millstone E. 2001. Mad cow disease — 1980s–2000: how reassurances undermined precaution. In: Gee D, Harremoes P, Keys J, MacGarvin M, Stirling A, Vaz S, Wynne B, editors, Late lesson from early warnings: the precautionary principle 1898–2000. Copenhagen: European Environment Agency, p. 157–167. http://reports.eea.europa.eu/environmental_issue_report_2001_22/en/issue-22-part-15.pdf. Accessed May 5, 2007.
von Winterfeldt DV, Edwards W. 1986. Decision analysis and behavioural research. Cambridge (UK): Cambridge University Press. 604 p.
Wallsten T. 1990. Measuring vague uncertainties and understanding their use in decision making. In: von Furstenberg GM, editor, Acting under uncertainty. Norwell (MA): Kluwer, p 377–398.
Winkler G. 1986. Necessity, chance and freedom. In: von Winterfeldt D, Edwards W, editors, Decision analysis and behavioural research. Cambridge (UK): Cambridge University Press.
Wynne B. 1992. Uncertainty and environmental learning: reconceiving science and policy in the preventive paradigm. Glob Environ Chang 2:111–127.
Zadeh LA, Kacprzyk J. 1992. Fuzzy logic for the management of uncertainty. New York (NY): Wiley. 676 p.

APPENDIX 4.1: GLOSSARY

Ecological footprint (EF): The total area of terrestrial and aquatic ecosystems required to fulfill all of the resource needs of a given human population, including waste assimilation.

Genuine progress indicator (GPI): An aggregated, monetized index that accounts for a broad array of sociocultural and environmental benefits and costs in the calculation of gross domestic product (GDP).

Green accounting: Explicit economic recognition of natural resource benefits and depletion and/or degradation costs in measures of economic health and performance such as the green GDP.

Green GDP: A **gross domestic product** (GDP) calculation that broadly accounts for the benefits of natural resource extraction and use as well as the costs of natural resource depletion and environmental impact.

APPENDIX 4.2: DECISION ANALYSIS TOOLS

Management decisions for complex ecosystems typically involve multiple decision makers, stakeholders, analysts, and other interested parties providing inputs to the decision-making process. This process can benefit from a definitive evaluation procedure that can be replicated, can be used to justify a decision, and can provide a clear rationale as to why specific decisions were made and who was involved in that process. Under ecosystem restoration and evaluation projects, there are typically nonmonetary outputs such as habitat units or acres of wetlands, which require monetization in formulating economic development or damage assessment plans. This has fostered an increased need for tools and guidance to conduct trade-off analyses and collaborative decision making.

TABLE 4.2
Decision-making procedure

Specify issues and opportunities: management goals	Statement of problems and opportunities
	Project scope
	Planning objectives and constraints
	Public, institutional, and technical significance of resources
	Stakeholder feedback
Inventory and forecast of conditions	Understanding of ecosystem structure and function
	Conceptual model of ecosystem that identifies key resources and processes
	Quantitative ecological model
	Range of variables considered (how to select useful and valid indicators and analyses)
Develop alternatives	Develop range of alternatives that achieve management objectives
Evaluation of effects	Assessments or differences between alternatives (including no action)
	Qualification or disqualification of alternative for or from further consideration
Comparison of alternatives	Comparison of differences between alternatives
	Trade-off analysis
Alternative selection	Select a proposed recommended alternative
	Obtain stakeholder feedback
	Public review
Overall process	Stakeholder awareness and involvement (including public)
	Interagency coordination
	Develop solid working relationships

Valuation Methods

Table 4.2 provides a proposed decision-making procedure starting from the development of the management objectives and goals to the selection of an alternative or series of alternatives that achieves the stated goals.

Making decisions involves making trade-offs. A trade-off means giving up one thing to gain another. Take the example of a reservoir storage reallocation. Will water be allocated to hydropower or flood control? Water can be withdrawn for irrigation, or water supply, or it can be used for in-stream uses such as navigation and habitat. In the case when land or water resources have competing and mutually exclusive uses, the trade-off is explicit (more of 1 unit value means less of another unit value) and physical laws may fix the terms of the trade-off. Implicit trade-offs, by contrast, are fixed by the value systems and preferences of decision makers and stakeholders. The trade-off is implicit since the terms of the trade are based on something other than physical laws, such as in the case of 2 alternatives, 1 of which will create a wooded area and the other a wetland area. Each interested party will have different reasons for wanting 1 or the other alternative, and these factor into the implicit trade-off.

There are a number of decision analytic methods and tools available to assist analysts and decision makers in synthesizing and integrating data and model results from the many different studies and analyses typically available for any given management context. These tools can be quantitative or qualitative in perspective. This appendix provides a brief overview of these methods and tools and their applicability to decision making.

A.1 DECISION ANALYTIC FRAMEWORKS

A.1.1 CONCEPTUAL APPROACHES

Here we provide an overview of approaches for decision analysis that are common in the resource management literature. These approaches are not discussed in detail but are outlined in terms of their potential for use in ecological risk assessment.

A.1.1.1 Stochastic Dynamic Programming

Stochastic dynamic programming is a mathematical programming approach that provides an optimum (a time path that characterizes the optimal strategy) given an objective function, a set of constraints, and information on the probability distributions of factors affecting the decision (see Miranda and Fackler 2002). The probability distributions typically describe different states of the world and are often used to construct a multistage decision problem where the decision maker is assumed to act, a random process occurs and affects the outcome, and then the decision maker acts again in response to the outcome of the random process (and so on). Stochastic dynamic programming provides insights into decision paths and the impact of varying degrees of risk on decision outcomes. It also provides information on the shadow prices of constraints that can help decision makers identify critical factors through time. The valuation approaches described above can help provide inputs into stochastic dynamic programming models, or the output from these models can provide information that can be used in a valuation exercise.

A.1.1.2 Bayesian Analysis

Ecological risk assessment almost always involves risk or uncertainty. Bayesian analysis is an approach to incorporate new information into the decision-making process (see Conrad and Clark 1987). Historical information (the prior) is updated by new information (observations or experiments), and of the new information affects the degree to which the historical perspective is modified (resulting in the posterior distribution). New information is a form of learning in the sense that the probability distributions characterizing the outcomes are improved (see Kinzig et al. 2003).

A.1.1.3 Multicriteria Analysis

Multicriteria analysis (or multiattribute decision analysis) is an approach to structuring decision problems. In its basic form, a single decision maker is assumed to be examining a decision problem with one or more attributes (elements that contribute to utility), a number of time periods, and a degree of uncertainty. The approach involves a systematic analysis of objectives, trade-offs, and risks. The formalization of utility weights, risks, and objectives is then typically evaluated in an optimizing framework (e.g., dynamic programming). There are many variants of multicriteria analysis depending on the specification of the problem. In general, however, it is a prescriptive approach for a single decision maker interested in a systematic approach to solving a complex decision problem (Keeney and Raiffa 1976). An example of a multicriteria framework, used in a more open and participatory mode to explore heterogeneity in valuations, is the multicriteria mapping approach (Davies et al. 2003).

A.1.1.4 Simulation Analysis

Simulation modeling for ecological risk assessment is based on the development of mathematical representations of ecosystem and human system processes (Maxwell and Costanza 1995). Simulation models can be very simple or very complex depending on the processes specified and the linkages between the equations and sectors. Simulation models do not provide optimal solutions in terms of optimization over an objective function. However, they have considerable flexibility in including a variety of components as variables in the system. Their use has increased with increases in computer processing speed and the development of graphical interfaces. They have been used as inputs to valuation processes, and valuation or behavioral information has been incorporated into such models as inputs (Maxwell and Costanza 1995).

A.1.2 APPLICATIONS OF DECISION ANALYTIC FRAMEWORKS

In addition to the conceptual approaches to decision analysis, several approaches specific to environmental risk management have been developed. These include some that are computer software–based templates for analysis. We review a selected sample of these frameworks below.

Valuation Methods

A.1.2.1 Soft Systems Methodologies

Checkland (1981) describes a "soft systems methodology" that aids the development of a narrative about the structures and processes interacting within a given environmental framework. One portion of Checkland's soft systems methodology is the CATWOE process. CATWOE is an acronym, the elements of which define the primary components of a system:

Customer(s): who benefits from the system, or who is affected by the system
Actor(s): those operating within the system who can effect transformation
Transformation: a change in state — usually phrased as "from X to Y"
Weltanschauung: worldview, typically pertaining to why a transformation is desired
Owner(s): who can shut down the system or change the transformation goal
Environment: external constraints on a system; factors outside of the actors' control

The process starts with a description of the perceived problem. How the problem is perceived by a group of stakeholders influences the CATWOE elements. Once the CATWOE elements are described, one can write a "root definition" that incorporates the various elements into a narrative, providing problem focus and solution guidance. The description of the problem as perceived and the root definition that derives from the process represent only one of many possible problem or root definition statements that can be made. The interesting aspect of the CATWOE process is that disparate groups of stakeholders can formulate different problem statements and root definitions, thus providing a focused version of their worldview. Stakeholder group members can then read the statements of other groups. Ideally, all participants then will come to a fuller understanding of the biases held and viewpoints taken of not only other groups but also their own.

A.1.2.2 Adaptive Methodology for Ecosystem Sustainability and Health (AMESH)

Kay et al. (1999) described a scheme to define context and constraints with a list of questions designed to elicit the operational aspects of complex adaptive systems. The questions are divided into sets, the answers to which evolve into a set of narratives painting a rich picture of system function and states. The 4 narratives included systems and issues of interest, material and energy flows, possible stable system states (basins of attraction), and causal loops and feedbacks present in the system. These methods have been refined and implemented in a number of areas. Working with people who reside within a system, adaptive methodology for ecosystem sustainability and health (AMESH) seeks to develop "influence diagrams" that highlight what the residents feel are relevant features and limitations, needs for and outputs from the village or community, and key institutions in the system. Often with no quantitative data requirement but drawing lines indicating decreasing or increasing resources and nuisances, an influence diagram can move the group to consensus on what is the primary "thing to fix" in addressing the identified village needs.

A.1.2.3 Ecosystem Management Decision Support (EMDS)

The Ecosystem Management Decision Support (EMDS) system integrates the NetWeaver knowledge base system with ArcView GIS to provide decision support technology for ecological landscape analysis applications. The NetWeaver technology that underlies EMDS enables more flexible problem-solving knowledge representations that permit an evaluation of the degree of truth rather than the binary true or false of more traditional rule-based approaches. Additionally, NetWeaver allows an assessment of the effect of missing information, and the incorporation of knowledge bases that can evaluate a range of topics, such as social, economic, aesthetic, and legal issues, which might be related to the ESC biophysical model.

5 Complexity in Ecological Systems

Katherine von Stackelberg, Elaine Dorward-King, Sam Luoma, Ron McCormick, Kristin Skrabis, and Stephen Polasky

CONTENTS

5.1 Introduction 98
5.2 Confronting Complexity, Dynamics, Variability, and Uncertainty in Ecosystems 100
    5.2.1 Seeking a Common Level of Analysis 100
    5.2.2 Scale and Ecological Type: Level of Analysis and Conceptual Models 102
    5.2.3 Variability and Uncertainty in Managing Complex Systems 103
    5.2.4 Trade-Off Analyses 106
    5.2.5 Summary: Systems Simple and Systems Complex 107
5.3 Integration of Ecology and Economics 107
5.4 An Example of an Ecologist and Economist Discussing Uncertain Valuation of Habitat Loss with Potential Extinction Risk 115
    5.4.1 Statement of the Problem 116
    5.4.2 Available Information 116
    5.4.3 Application of Benefit–Cost Analysis 116
    5.4.4 Uncertain Future Losses from Extinction 117
    5.4.5 Additional Considerations 118
5.5 Case Study: Extractive Development and the Valuation of Biodiversity and Community Needs in Madagascar 119
    5.5.1 Description of Complexities 119
    5.5.2 The Natural Environment 119
    5.5.3 People with Relevant Interests 120
    5.5.4 Integrated Solutions 121
    5.5.5 Significant Outcomes and Lessons Learned 121
5.6 Complexity of the System: CALFED Case Study 122
5.7 PCB Case Study 124
5.8 Lessons Learned and Not Learned 125
    5.8.1 Integrating Ecology with Economics 125
    5.8.2 Appreciating and Understanding Complexity 125
References 127

5.1 INTRODUCTION

> Complex: consisting of parts or elements not simply coordinated, but some of them involved in various degrees of subordination; complicated, involved, intertwined, intricate; not easily analysed or disentangled.
>
> *Oxford English Dictionary*

Complexity is inherent in ecological systems due to the self-organizing, deeply hierarchical nature of energetically open systems (Kay et al. 1999). Humans attempt to manage ecological systems by organizing and evaluating individual components of the system, typically by using mechanistic or other kinds of models that describe, and ideally quantify, specific physical processes. The goal is to capture physical processes and interactions in sufficient detail that the primary mechanisms for responding to changes are quantified, but not so detailed that parameterization and interaction of parameters become intractable. Ecological systems are dynamic and adaptive, and much as we try to understand or model systems from first principles, we have difficulties implementing effective organizational learning in our management of these systems.

Decisions are required every day for determining how ecological resources will be explored, used, and managed by humans. But decision makers and analysts always have an imperfect understanding of how such systems function and evolve. Much as complexity is inherent in all ecological systems, uncertainty is inherent in every environmental policy decision, whether it is acknowledged or not. In addition, there can be multiple, potentially conflicting goals for the use of natural resources or for what are the "best" uses for an ecosystem and the habitats it may contain. Resiliency and evolutionary adaptation in nature are particularly difficult to understand and model, and represent a fundamental aspect of the complexity we observe in our attempts to understand ecological systems.

Nonetheless, imperfect knowledge notwithstanding, we must make decisions regarding the use of environmental resources and the attendant impacts to ecological systems. Management of these systems can range from fine-scale, localized decisions with relatively short-term and temporary effects, to broad-scale decisions with long-lived regional or global impacts. Impacts from changes to ecological systems affect many different segments of society and have costs and benefits associated with them that may be unequally distributed across different stakeholders (e.g., 1 group may receive disproportionate benefits relative to costs than another). All decisions will affect a suite of stakeholders with multiple and potentially conflicting management objectives. We have great difficulty in defining the full consequence of proposed changes to ecological systems, and in fully understanding those changes in hindsight. "Wicked problems" are defined as those for which decision making involves multiple stakeholders with multiple objectives, and sizable uncertainty exists about the effects of certain actions proposed to meet management objectives (Rittel and Weber 1973). By these criteria, every decision involving ecological systems is a "wicked problem."

Ecological risk assessment and valuation of ecological systems aim to provide decision makers with the best available information about the relative merits of alternative courses of action. Acknowledging imperfect understanding and limited ability to accurately predict consequences, having a greater understanding of system dynamics, and possessing better information provide decision makers a greater chance of realizing beneficial outcomes. Well-informed and sustainable decisions need information about likely consequences of actions. The ability to supply such information requires analyses that adequately address system complexity. By definition, such analyses require inclusion of multiple spatial and temporal dimensions and effective integration of input from multiple disciplines. Whether it is a private sector decision to embark on a potentially controversial multimillion dollar expansion project, or a judgment by USEPA on the remediation approach for a large contaminated site, each requires a deep and thorough understanding of the complexity and the uncertainties involved. This will always require a model, whether quantitative (e.g., defined mathematical relationships between components), qualitative (e.g., graphical or narrative descriptions of interactions), or, in most cases, both.

However, a model cannot by itself capture complexity — data, monitoring, and observations are required to inform the modeling process. There is greater predictive power in models that are constructed from first principles — that is, in which the representation of physical processes and interactions is based on established theories, ideally verified through a combination of experimentation and observation. We have greater predictive power if we can quantify how and why a system responds to changes. But that predictive power is compromised in the absence of information specific to the management decision. For example, bioaccumulation modeling of hydrophobic organic contaminants in aquatic systems is well rooted in theory and experimentation. But a bioaccumulation model applied to a system for which there is no site-specific data for key parameters is so uncertain as to be virtually useless in terms of decision making. The other issue is that there may be multiple models that satisfy scientific criteria without an obvious way to determine which one is preferable.

Decisions that affect ecological systems are made at many technical, managerial, and leadership levels in all sectors of society. The degree of detail to which complexity must be expressed and captured depends upon the level of the decision maker and the magnitude of the decision. Decision makers at any given level will have specific requirements depending on the scope, time frame, and long-term implications of the proposed management actions. Analysts must always provide justification for the model or suite of models that are selected or developed to provide decision makers the technical information necessary to make a decision. Equally important are the monitoring and specific environmental measurements that are used to develop an understanding of ecosystem structure and function. Models are essential for ensuring that complex systems are parameterized with the goal of understanding, protecting, and managing in a manner consistent with the values of the relevant stakeholders. But they are close to useless without data to inform their application. Finally, political pressure simplistically applied to obfuscate complexities increases the risk of insufficient rigor to implement an effective policy.

The next section of this chapter contains a primer on important aspects to consider when describing complexity in ecological systems. Section 5.3 discusses alternative decision analysis tools for incorporating information from disparate sources into an integrated analysis of complex systems. Section 5.4 describes a set of case studies that illustrate important issues involved in ecological risk assessment and system valuation. Section 5.5 summarizes lessons learned, and not learned, to date from the practices of ecosystem risk assessment and system valuation for decision making.

5.2 CONFRONTING COMPLEXITY, DYNAMICS, VARIABILITY, AND UNCERTAINTY IN ECOSYSTEMS

5.2.1 SEEKING A COMMON LEVEL OF ANALYSIS

> One should not increase, beyond what is necessary, the number of entities required to explain anything.
>
> **Occam's Razor**

It is self-evident that biological and/or ecological systems are complex. Complexity does not mean that orderly analyses of such systems are futile. Effective description of a system can constrain complexity to manageable dimensions. In our efforts to explain complexity to multiple scientific disciplines, we need to understand the importance of any constraints we impose on a system.

Humans have come to recognize the complexity inherent in ecological systems (Kay et al. 1999). However, the hierarchical nature of complex systems makes them inherently well ordered, and thus describable (Ahl and Allen 1996). The challenge to interdisciplinary analysts lies in describing that order from a common viewpoint, so that observers from different perspectives can effectively communicate. The standard approach to describing ecological systems involves ordering them from cell to organism to population to community to ecosystem to landscape to biome, with the implicit assumption of a systematic increase in scale with each new level. Although such scaling relationships represent a contrivance on the part of the observer (Allen and Hoekstra 1992), they do order the components of the system such that an entity chosen by the observer will stand out from its background.

Component processes and structures of a system work together in a defined, scale-dependent order. Physiology connects organs to bound an organism. Spatial relationships defined by the observer connect patches of vegetation to bound a larger landscape. Issues of spatial and temporal scale extend beyond their implicit association with the ordered system components. A population, by definition, is composed of 2 or more similar organisms, but spatial scale does not necessarily bound a population; the observer's decision to include or exclude individuals does. However, a focal spatial scale defining the problem area (riffle, run, river, wetland, watershed, etc.), and recognition of connecting processes, is essential to effective interdisciplinary communication and, hence, analysis. In many cases, spatial boundaries for a particular analysis are established by property ownership and legal liability rather than from an understanding of ecosystem dynamics. Of course, populations of animals do not recognize such boundaries in the absence of physical constraints to their movement.

Hierarchy theory (Ahl and Allen 1996) presents an approach to addressing complexity that provides a framework for differentiating an object from its context and for providing orderly cross-scale translations. It provides for a unified approach when multiple disciplines confront the same ecosystem complexity. Characteristic processes, rates of exchange within and among processes, physical scale, and ecological type (e.g., watershed) differentiate levels of analysis. Consider a leaf on a tree: the leaf is part of a larger organism, a tree, which itself is part of a larger community, a forest. Photosynthesis and respiration are biochemical processes occurring within the leaf. The forest provides the upper-level context for the tree. The tree provides a context for the leaf (i.e., a place to hang from and a supply of water and other elements), and the leaf provides a context (i.e., cells, gas, and material transport) within which photosynthesis and respiration occur. Forest communities develop over long time periods, and typically persist much longer than any individual tree. The tree long outlives the leaf, and a single leaf experiences numerous chemical cycles. Each level exhibits particular histories, rates, and behaviors that help differentiate observational foreground from background; but each is also dependent upon its larger and smaller context. Physical scale is similar. Analytical scale increases commensurate with ecological type, that is, cells within a leaf, a leaf on a tree, a population of trees within a forest community, a forest community on a landscape, and an animal-groomed landscape that is part of a biome.

From the observer's point of view, there are always at least 3 levels in any hierarchy: the level of interest, a lower level, and an upper level (Ahl and Allen 1996). The scale of observation and organizational level of interest differ with the observer. From the viewpoint of an insect (or an entomologist, for that matter), the worldview can scale down to cells and up to a leafy landscape (not down to photosynthesis or up to a tree and forest). So it is useful when setting up interdisciplinary studies to acknowledge that scale is observer independent, even though some argue for a more fixed state defined by terms such as "natural" and "theoretical" scale (e.g., Baveye and Boast 1999). A mutual agreement on "level of analysis" is key to effective analyses of complex systems, especially between people from disciplines presumably as different as ecology and economics. A common level of analysis serves to define more precisely *why* the problem appears complex and thus *what* elements we perceive as important to the system.

Developing a narrative describing essential structure–process interactions establishes a base-level model of system behavior. Only from such a qualitative framework can more quantitative models arise. Typically, structural models of a complex system use a mathematical framework that describes physical processes at a level of detail commensurate with the quantitative data available within the context of the management goal. The quantitative model further informs decision makers as to what aspects of the system are important to measure. There can be a large number of possible models to explain a given set of observations or data. This is because a model can represent an infinite number of possible cases, of which the observations are only a finite subset. The cases we haven't observed, or responses of the system to specific changes, are inferred by postulating general rules addressing both actual and potential observations. This is complicated by the fact that there are no strict scientific criteria for maintaining or restoring an ecosystem. Ecosystems are, by definition, dynamic and adaptive, and it is a value judgment as to which ecosystem is "preferable" to another (e.g., the creation of wetlands).

5.2.2 Scale and Ecological Type: Level of Analysis and Conceptual Models

> In terms of conventional physics, the grouse represents only a millionth of either the mass or the energy of an acre. Yet, subtract the grouse, and whole thing is dead.
>
> **Aldo Leopold,** *Sand County Almanac* (1948)

Most modern ecologists acknowledge that an explicit conceptual model is critical to identifying the components and relationships of an ecosystem in a manner useful for interdisciplinary analyses. In all cases, the observer must explicitly consider scale and type (Allen and Hoekstra 1992). Selecting an appropriate scale sets the "level of observation" and establishes the boundaries of the system in question. Selecting an ecological type sets the "level of organization" and establishes that which will be observed within the bounded system.

Level of organization + Level of observation = Level of analysis

Explicit definition of level of analysis is the first step to constructing a useful conceptual model.

Scale, whether temporal or spatial, decomposes into 2 component elements, namely, grain, the finest resolvable unit of interest, and extent, the widest range over which the units of interest are observed. Grain and extent set the lower and upper bounds, respectively, of what is distinguishable analytically. For example, making water quality measurements on a monthly basis (grain) does not inform what happens on a daily basis, but may be sufficient to define annual trends (extent). In setting grain and extent, the observer sets up several implicit and explicit relationships between each observation or measurement, including how a measurement is to be made, the time and space interval between measurements, how the measurements will be integrated (e.g., summed, averaged, ranked, and grouped), the overall spatiotemporal extent that observations are made, and a definition of what constitutes a difference in measured values (Allen et al. 1987).

Most observers focus their analyses on 1 of 6 ecological types (i.e., organism, population, community, ecosystem, landscape, or biome). Organism and landscape are types that are relatively recognizable by any observer, but population, community, and ecosystem are more abstract types and not as easily bounded. Once type is defined, the observer has to make an arbitrary decision as to observational boundaries. If the type is an individual human, this seems simple. However, consider whether the communities of mites covering our skin, or the sweat, pheromones, and other biochemical by-products we exhale or exude from our skin, are appropriate to the bounding definition of "individual organism." Or, when studying a population of fish, do we include all members of a species scattered throughout the world, just those in the United States, just those in 1 river in the United States or just those occupying 1 reach of 1 river in the United States? Further, do we include all the genetic material from all fish in this 1 particular river? Observational grain and extent must be compatible with one another, if the study is to provide useful knowledge.

In the fish example, the more rivers and species included, the less detail we can tractably consider about each individual fish unless some other aspect of the problem is constrained.

The world is too complex for a single scale to be associated with a given type. Landscapes are a defined spatial extent, the bounding of which is completely dependent on the observer, that is, there is no "landscape scale" (Allen 1998). A watershed represents a specific type of landscape defined by a single process: in simplistic terms, the direction rainfall runs when it hits the ground. But even for landscapes bounded by this simple process, the spatial extent can vary greatly. The Mississippi River Basin has a defined watershed boundary, as does a small creek. A typical study might then confront complexity by constraining data collection to third-order (grain) watersheds (type) within the Mississippi River Basin (extent). It is possible to aggregate data to higher order watersheds, but not lower. Constraints are always essential to tractability. Explicit acknowledgment of scale (grain and extent) and type sets the common level of analysis, clearly specifies how the observer is interacting with the system observed, and addresses complexity in an orderly manner.

Those observer-based decisions regarding the conceptual model of a system can be driven by practical considerations (available tools and data), by management goals, or by the research paradigm within which they work. Nevertheless, explicitly defining the relationship between components of any hierarchy is imperative. As pointed out, the explicit decisions researchers make in selecting organizational type of interest and observational grain or extent will define the level of analysis for addressing one specific problem. Changing grain or extent changes the level of analysis, and thus addresses a specific problem that is different from the original, not just larger or smaller.

5.2.3 VARIABILITY AND UNCERTAINTY IN MANAGING COMPLEX SYSTEMS

What the physicist views as noise is music to the ears of the ecologist.

Unpublished quote from a 1998 Pellston conference on ecological variability

For economists and ecologists to work together in valuing ecosystem services, it becomes essential that both understand and appreciate the complexity of quantifying ecological attributes. This section describes how uncertainty and variability might affect attempts to quantify a resource issue, and how successful ecological studies can deal with or even exploit environmental variability. We intend to offer help to economists and ecologists in better understanding the limits and capabilities of the science that underlies an effective resource valuation of a complex ecosystem. Our ultimate goal is to help study designers avoid the misrepresentations that often accompany oversimplifications in ecosystem definitions.

Environmental stochasticity poses one of the great challenges to understanding ecosystems. While many properties of ecosystem variability are difficult to predict precisely, all environmental variability is not random. Properties that seem to vary stochastically at a fine scale may, at a coarser scale, vary in a repeated, predictable fashion (i.e., variability is nondirectional). These are called "dynamic stabilities." Ecological structures and processes evolved in environments replete with dynamic stabilities. The dynamic stabilities of physical and chemical cycles, for example, are incorporated into the adaptations that drive the life histories of all species. Both environmental stochasticity and dynamic stability add complexity to properties, fluxes, and rates in an ecosystem. However, because dynamic stabilities involve somewhat predictable patterns, they also offer opportunities to understand those complexities, once the patterns are recognized.

Recognizing how variability affects the measure (or value) of a resource is the first step to appreciating both the limits and the opportunities offered by variability. Abiotic environmental processes (climate, hydrology, physical characteristics, and biogeochemical cycles) all exhibit an inherent variability. This variability affects the exposure of system elements to stressors, and affects our ability to detect and link system response to the stressor. Additionally, variability in the stressor gets superimposed on a fluctuating and nonlinear biotic response. For example, in analyzing pesticide use and effects, a first consideration is variability in use and movement into the environment (e.g., pesticide movement into streams may be delayed during seasons of low rainfall). A second consideration is what species and life stages are present at the time pesticides are transported. We may estimate the dose that causes toxicity to a resource species from an experiment. But in nature, dose (exposure) may vary by orders of magnitude with season, and sensitivity may vary with life stage. The presence of elevated pesticide and sensitive life stage must coincide for maximum sensitivity. Evaluations of costs of a pesticide effect can be completely in error if such simple dynamics are not considered.

Experimental studies isolate and measure a single variable and control all others. They seek to reduce and control variability of the environment in an effort to confirm a single hypothesis. Experiments are necessary to plausibly demonstrate mechanisms, but the full complexity of multiple, interacting variables is difficult to study experimentally. Inevitably, critical feedback loops are lost (Cairns and Mount 1990). An alternative is field observation and field monitoring. These "observational" studies are vital to achieving a better understanding of dynamic systems. Long-term (more than a decade in duration) studies of the same ecological system are an effective way of progressively understanding complexities that appear intractable in simpler studies (Cloern 2000; Hornberger et al. 2000). Nevertheless, cause and effect can only be circumstantially linked in the field; unequivocal causal inference is rare. So both field and laboratory science have inherent imperfections. Combining experimentation, manipulation, and observation is effective in enhancing understanding. But ecological interpretations and economic valuations need to appreciate the inherent limits of the results from the different approaches and synthesize different lines of evidence within those limits.

Misrepresentations can occur if analyses (e.g., trade-off analyses or valuations) do not adequately address variability (Luoma 1996). But there are effective approaches to avoid such pitfalls. One example is scenario building and forecasting. Forecasts can evaluate outcomes when different environmental circumstances are built into scenarios of the way systems might work (Presser and Luoma 2005). Such approaches provide opportunities to incorporate economic analyses with the consideration of ecological outcomes from the different scenarios.

Luoma et al. (2001, p. 141) offer the following as a summary of the consequences of ecosystem variability:

> Variability is a property of complex natural systems that defines dynamics and influences uncertainty. Variability is both an impediment to clear understanding and part of the richness of natural systems. Variability that is not characterized or accounted for (excessive simplification of processes) leads to uncertainty in the results, and even to complete masking of important outcomes.... Our central thesis here is that environmental variability is not intractable, but that with judicious design, analysis, and interpretation, it can be characterized and controlled, so that the uncertainty of field studies and field experiments can be reduced and interpretations optimized.

Uncertainty and variability should be viewed separately in any analysis because they have different implications to regulators and decision makers (Morgan and Henrion 1990; Thompson and Graham 1996; von Stackelberg et al. 2002). Variability is a population measure (although individuals do display temporal variability in body weight, exposure, etc.) and is intrinsic to the system. Variability typically cannot be reduced, only better characterized and understood. In contrast, uncertainty represents unknown but often measurable quantities. Typically, obtaining additional measurements of the uncertain quantity can reduce uncertainty. Risk assessment provides an example of the utility of operationally separating uncertainty and variability. Quantitatively separating uncertainty and variability allows an analyst to determine the fractile of the population for which a specified risk occurs and the uncertainty bounds or confidence interval around that predicted risk. Disaggregating sources of uncertainty and variability within a decision framework assists decision makers in identifying areas of potential research for uncertainty reduction versus identifying conditions under which numeric predictions cannot be further reduced because they are attributable to natural variability in parameters.

Uncertainty is the result of our inability to fully quantify how a dynamic system functions and how the components interact. If uncertainty is large relative to variability (i.e., contributes most to the range of risk estimates) and if the consequences of making the wrong decision are great, then additional collection and evaluation of information are necessary before making management decisions. Uncertainty can be reduced through systematic field and laboratory studies to inform a better understanding of biotic–abiotic relationships and can be minimized through appropriate and rigorous attention to study design and execution. Difficulties exist in quantitatively partitioning the source and effect of uncertainty, as well as in deciding what to do about any uncertainty one does quantify. In the end, however, decisions require a choice as to the acceptability of different types and magnitudes of uncertainty. The types of decision error that uncertainty leads to, and the burden of proof that emerges, should be recognized in any intertwining of ecology and economics.

Two principle error types result from uncertainty. A type I error occurs when an observer incorrectly concludes that an effect occurs, when in fact there is no actual effect, just random errors creating the perception of an effect. To avoid type I error, observers search for evidence that unambiguously demonstrates an effect. Only when this evidence exists can the observer draw conclusions. In instances where avoiding a type I error is emphasized, the burden of proof is placed on those asserting the existence of an adverse effect and a link to the suspected cause. The implication can be that no effect is attributable to that cause unless the linkage is demonstrated. Awareness of type I and type II errors is just as important in studies that intertwine economics and ecology as it is for the ecologist studying effects of stressors in nature.

Seeking to minimize type I error increases the risk of committing a type II error. Type II errors occur when the observer concludes "no effect" when, in fact, an effect has occurred. Such outcomes are most common when available methodologies are not sufficiently sensitive to detect the effect. Environmental variability reduces the sensitivity of field studies, resulting in a larger risk of type II errors than in a controlled experiment. Presently, our overreliance on traditional, statistically based hypothesis testing lies at the core of many type I–type II error dilemmas. In formal terms, the only allowable conclusion from a failure to demonstrate an effect is that no conclusion can be drawn. Increasing the certainty that an effect is unlikely is best accomplished by using a sufficient number of rigorous studies and addressing the problem using several different approaches to establish concordance. Model estimation, maximum likelihood methods, or Bayesian statistics offer alternatives to traditional statistical approaches for both ecologists and economists. There is a formal literature defining how feasibility of an effect should be evaluated (e.g., Brown et al. 2002).

Analysts and decision makers are inevitably confronted with questions about how uncertainty, natural variation, bias, and precision in the input affect conclusions drawn from risk assessments and other studies. For example, when assessing risks to endangered species, analysts may face questions as to whether the breadth of variation in exposure on an hourly, daily, or yearly basis has been adequately characterized and quantified. In the same assessment, substantively different but related questions might also be raised regarding the uncertainty of the quantitative estimates used. In other settings — when assessing risks to populations, for example — variation in exposure between individuals can become a key issue, and along with it the ubiquitous need to assess either bias or precision in estimating this variation. Similar concerns may be associated with other inputs into the risk assessment model, such as toxicity estimates. Uncertainty and variability are inherent in our understanding of complex systems, and approaches exist for quantifying and parsing their influence on results of analyses.

5.2.4 Trade-Off Analyses

Management decisions for complex ecosystems typically involve multiple decision makers, stakeholders, analysts, and other interested parties providing inputs to the decision-making process. This process can benefit from a definitive evaluation procedure that can be replicated, be used to justify a decision, and provide a clear rationale as to why specific decisions were made and who was involved in that process. Situations with considerable uncertainty and variability, as has been discussed here, lend themselves particularly well to trade-off analysis. Under ecosystem restoration and evaluation projects, there are typically nonmonetary outputs such as habitat units or acres of wetlands, which require monetization or some type of valuation in formulating economic development or damage assessment plans. This has fostered an increased need for tools and guidance to conduct trade-off analysis and collaborative

Complexity in Ecological Systems

decision making. Finally, the fact that there is typically no scientific basis alone for making a decision requires tools that allow for trade-offs based on expected outcomes under different management alternatives and the values that stakeholders place on those outcomes. Chapter 4 provides a description of the specific tools that can be used to conduct trade-off analyses (e.g., multicriteria decision analysis).

5.2.5 Summary: Systems Simple and Systems Complex

Everything should be as simple as possible, but no simpler.

Albert Einstein

Normal science advances by defining a system into simplicity (Kay et al. 1999) so problems are tractable and the role of observer can be ignored. With simple systems, everyone agrees as to what is structure versus behavior, what changes are discrete versus continuous, and what differences are qualitative versus quantitative. No such unanimous consensus exists for complex systems. To deal with complex systems, researchers set a common level of analysis, establishing explicit definitions regarding the significance and meaning of system components. The system itself will not tell the researcher how to make these decisions, so the researcher must accept responsibility for those judgments and cannot pretend to be objective.

Simplicity or complexity is not a characteristic of the material system itself, but is merely one possible expression of how the researcher chooses to cast the system. In evaluating complex systems, it is important to note the following:

- Environmental variability is not just "pervasive stochastic variability," error, or measurement of experimental error.
- Dynamic stabilities (properties that vary in a repeated, somewhat predictable fashion) add recognizable order to complex systems.
- Appreciating and capturing the nature and the role of environmental variability are critical to avoiding oversimplifications that could misrepresent ecosystems, quantifications of their resources, and evaluations of trade-offs among issues relevant to those systems.
- There are more sophisticated, quantitative methods available (e.g., Bayesian statistics) for incorporating and maximizing the information obtained from highly variable systems and for explicitly accounting for the relativism inherent in any evaluation of ecological systems.

5.3 INTEGRATION OF ECOLOGY AND ECONOMICS

The primary issue for integration is that economists need to establish the value of an ecological resource based on sound science, usually within a regulatory framework and often based on stakeholder preferences. The objective of this section is to explore the types of data and model results typically produced from ecological studies in relation to those data needed for economic valuation. Table 5.1 describes

TABLE 5.1
Integration of ecology and economics

Resources	Types of scientific analysis	Types of outputs	Potential for integration			Notes
			Low[a]	Medium[b]	High[c]	
Aquatic						
Drinking water	Concentration of contaminant(s) to identify exceedances of water-quality criteria, dissolved oxygen, pH, turbidity, and total suspended solids	Point estimates relative to regulatory limits (exceed a water-quality criterion)			√	Straightforward to estimate this commodity
Groundwater not used for drinking water	Concentration of contaminant(s)	Point estimates relative to regulatory limits.	√			Concentrations do not indicate how the in situ groundwater services may be impaired. More controversial when there is no currently planned use for an aquifer. Traditional economics do not readily recognize a value beyond the option to use in the future.
	Hydrologic studies	Volume injured to date, annual recharge rate, and porosity			√	
Surface water not used for drinking water	Concentration of contaminant(s) to identify exceedances for species and habitat health	Point estimates relative to regulatory limits	√	√		Exceedances can be incorporated into an index for use in economic analysis
	Recreational use of surface water				√	Can use established methods (e.g., travel cost) to value recreational use

Complexity in Ecological Systems

	Hydrologic studies	Variations in the hydrograph due to hydropower peaking operations lead to erosion (reported in river-miles or acres), losses from abnormal flooding patterns (e.g., acres of lost forest because of prolonged inundation), and effects of dam removal (identification of run of river).		✓	Direct inputs for evaluating the cumulative impacts of hydropower operations as compared to the power benefits
Terrestrial					
Soil	Concentration of contaminant(s)	Point estimates, distributions	✓		Need to know how the soil relates to other resources and the total losses resulting from impairment
	Bioavailability studies	Incorporated into estimates of exposure	✓		
	Soil invertebrate concentrations	Point estimates, distributions	✓		Cannot use concentration data directly
	Relationship between soil and invert concentrations	Ratio of invertebrate concentration to soil concentration		✓	
Vegetation	Concentration of contaminant(s)	Point estimates or distributions	✓		With the proper inputs, injury is easily valued using habitat equivalency analysis. Can also use survey methods to value willingness to pay for loss of habitat to ecological receptors based on existence value, bequest value, and so on.
	Site-specific relationships between soil and vegetation concentrations	Ratio of vegetation concentration to soil concentration		✓	

(*continued*)

TABLE 5.1 (CONTINUED)
Integration of ecology and economics

Resources	Types of scientific analysis	Types of outputs	Low[a]	Medium[b]	High[c]	Notes
	Vegetation and/or tree surveys	Types of resources by numbers of individuals and/or total area			✓	
	Invasive species	Types of resources by numbers of individuals and/or total area			✓	
	Habitat suitability evaluation	Habitat area and qualities, which may be indexed to identify percent loss of services		✓		
Wetlands	Habitat suitability evaluation	Habitat area and qualities, which may be indexed to identify percent loss of services			✓	Habitat equivalency analysis can be used for wetlands. Straightforward calculations. Better monitoring data on the functional capacity of created and restored wetlands would make economic analysis more defensible.
Sediment	Concentrations of contaminant(s)		✓			In practice, a limited understanding about the pathways from sediments and how sediment exposures influence potentially resulting injuries (although can evaluate injury to fish, birds, and mammals — resource users of sediments — directly).
	Total organic carbon	Fraction	✓			
	Measures of bioavailability	Influences exposure potential	✓			

	Measurement	Description	✓	Notes
	Concentrations of contaminants in pore water	Point estimates or distributions	✓	
	Redox potential	Influences exposure potential	✓	
	Grain size distribution	May influence exposure; influences habitat suitability	✓	
	Species			
Invertebrate	Concentrations of contaminant(s)	Mass contaminant per mass invertebrate	✓	
	Numerous indices (e.g., abundance, diversity, and Shannon's Index)	Index	✓	Index can be used for a habitat equivalency analysis.
Vertebrate — bird	Concentrations of contaminant(s)		✓	
	Bird surveys	Number of individuals, age distribution	✓	Direct inputs for recreational studies and resource equivalency analysis
	Habitat availability	Habitat types and area for foraging and reproduction	✓	Provides data to scale restoration
	Risk assessment results	Deterministic toxicity quotients (TQ); probabilistic (e.g., probability of effect or increasing magnitude of effect)	✓	Probabilistic outputs may be incorporated into an index to identify the percent loss of services; probabilistic results can be used with survey methods to obtain willingness to pay values for risk reductions.
	Population modeling (birth, mortality)	Annual survival rate by age, average life span, percent female, percentage reproducing females, and fecundity	✓	Direct inputs for resource equivalency analysis
	Egg exposure studies	Effects on reproduction		Can help inform potential for recovery

(*continued*)

TABLE 5.1 (CONTINUED)
Integration of ecology and economics

Resources	Types of scientific analysis	Types of outputs	Potential for integration			Notes
			Low[a]	Medium[b]	High[c]	
Vertebrate — mammal	Concentrations of contaminant(s)	Site-specific dose–response studies		√		Can be incorporated into an index to identify the percent loss of services
	Wildlife surveys	Number of individuals, age distribution	√			Direct inputs for recreational studies and resource equivalency analysis
	Habitat availability	Habitat types and area for foraging and reproduction			√	Provides data to scale restoration
	Risk assessment results	Deterministic toxicity quotients (TQ; probabilistic (e.g., probability of effect or increasing magnitude of effect)	√		√	Probabilistic outputs may be incorporated into an index to identify the percent loss of services; probabilistic results can be used with survey methods to obtain willingness to pay values for risk reductions.
	Population modeling (birth, mortality)	Annual survival rate by age, average life span, percent female, percent reproducing females, and fecundity			√	Direct inputs for resource equivalency analysis
	Site-specific dose–response studies	No observed adverse effect levels (NOELs), lowest observed adverse effect level (LOELs), and continuous concentration–effect curves		√		Can be incorporated into an index to identify the percent loss of services; NOELs and LOELs can be used to bound the range for mortality in resource equivalency analysis.

Complexity in Ecological Systems

	Type of study	Description	a	b	c	Use in economic analysis
Fish	Concentration of contaminant(s)	Exceedances used for fish consumption advisories (FCAs); NOELs and LOELs help bound the range for mortality	√			Exceedances used FCAs to help inform recreational studies.
	Creel surveys	Number of fish, age distribution, and fishing pressure			√	Direct inputs for recreational studies and resource equivalency analysis
	Habitat availability studies	Habitat types and area for feeding and reproduction			√	Provides data to scale restoration
	Risk assessment results	Deterministic toxicity quotients (TQ); probabilistic (e.g., probability of effect or increasing magnitude of effect)		√	√	Probabilistic outputs may be incorporated into an index to identify the percent loss of services; probabilistic results can be used with survey methods to obtain willingness to pay values for risk reductions.
	Population modeling (birth, mortality)	Annual survival rate by age, average life span, number reproducing, and eggs per unit (e.g., per female, salmon redd)			√	Direct inputs for resource equivalency analysis
	Site-specific dose–response studies			√		Can be incorporated into an index to identify the percent loss of services

[a] Low: although the outputs may inform other scientific studies, they cannot be used directly in economic studies.
[b] Medium: the output can be incorporated into a benchmark or translated for use in economic analysis.
[c] High: the output may be directly used in economic analysis without further adjustments.

Note: In some cases, 2 checkmarks are used to indicate there is a range in potential for integration.

information generated by typical risk assessment or environmental impact studies. Each element in the table is further evaluated as to how suitable (low, medium, high) these results are for subsequent economic valuation exercises. A low score indicates the information generated is not suitable for economic valuation purposes. Medium indicates that there is some potential for integration, but the data units are not commensurate with the units required for economic valuation. A high score indicates that the data occur in a form directly useful for economic valuation purposes.

In practice, the outcomes of risk assessment and other scientific studies are handed unfiltered to economists for ecosystem valuation. This practice results in a disconnect for decision makers in terms of planning with limited resources, communicating across disciplines, and transferring knowledge in a meaningful way so as to create an efficiency of effort and outcomes. How many times has a project manager thought, "If only they had measured Y instead of X?" Efforts to construct assessment points and benchmarks for economic valuation are a source of uncertainty, if not outright contention. How do varying levels of contamination in sediments relate to service losses in a river?

As shown in Table 5.1, only a handful of scientific analyses actually have a "High" likelihood of producing data directly integrated into an economic analysis. A larger number of analyses (denoted as "Medium") produce outputs that require adjustment or translation prior to use by economists. The majority of scientific analyses, though, produce point estimates that do not lend themselves to quantification of resource injury or measurement of changes in resources from different actions (integration is "Low").

Table 5.1 is not a comprehensive list of all scientific methods for evaluating different types of resources. Rather, the examples provided highlight the most common types of scientific analyses found in the literature and in practice. Further, the implied interface with economic tools is based on common types of economic analyses like travel cost and habitat equivalency. Some common areas of integration are revealed. In the case of natural resource damage assessment and restoration (NRDAR), for example, all services should be identified relative to the baseline conditions of the injured resources but for the release of hazardous substances. This is typically done by estimating percent loss of services. A variety of parameters provided by or adapted from scientific studies can be used to estimate percent loss, such as the following:

Mortality; reproductive rate; growth rate: Where there are dead or destroyed natural resources, the injury side of a Habitat Equivalency Analysis (HEA) or Resource Equivalency Analysis (REA) is less speculative (100% loss of baseline). Analysts strive to select relevant, feasible restoration projects to help ensure that the scaling is as accurate as possible. The restoration literature is more limited than the injury side, but does offer data on reproductive patterns and growth rates for natural resources, which serve as proxies for valuing return of services.

Bioaccumulation factors: Where more complex ecosystem injury occurs, analysts often identify appropriate "representative" species, communities, and/or habitat types that are, or directly support, injured trust resources for potential scaling. Bioaccumulation factors are sometimes used to identify potential injury to trust resources, through a particular pathway, for translation into service loss percentages. Care needs to be taken to ensure that exposure data, injury data, and/or literature values are consistent with the results of any modeling effort. In addition, analysts generally avoid "double counting" (i.e., overcounting) the extent of injury when selecting representative species with potentially overlapping food chains and multiple pathways of exposure.

No observed and lowest observed effect level (NOEL and LOEL): In the absence of on-site injury data, literature may provide qualitative indicators of injury as it relates to various concentrations of a chemical to bound assumed levels of injury. The extreme values of exceedances may be particularly reliable as measures of service loss percentages. For example, when < 1.6 PCB concentration (ppm net weight) occurs in certain fish, there are no observed effects on spawning, hatching, growth, and mortality (Bengtsson 1980). Alternatively, > 290 PCB concentration (ppm net weight) in certain fish leads to mortality (Mauck et al. 1978). Although limited, some studies provide empirical evidence of how exceedances of criteria can lead to injury (see, e.g., Long et al. 1998).

Established indices: Analysts consider ways to benchmark the service loss percentages to concentration ranges for both injury and literature values. An established and widely used index like the Habitat Suitability Index (HSI) may be useful.

The risk assessment process does not always generate the kind of data and results useful to quantifying and documenting service losses, although the framework for doing so exists. In some cases, there are more institutional barriers to effective integration of economics and science due to budgetary constraints, program mandates, and legal issues. Although in many cases it is technically feasible to collect and interpret data in ways relevant to achieving numerous objectives, in fact there exist regulatory hindrances to actually doing that. For example, under existing Food and Drug Administration regulations, human carcinogens must be managed at the 1 in a million (1×10^{-6}) risk level. This is the so-called "Delaney clause" that has been in effect for many years.

5.4 AN EXAMPLE OF AN ECOLOGIST AND ECONOMIST DISCUSSING UNCERTAIN VALUATION OF HABITAT LOSS WITH POTENTIAL EXTINCTION RISK

Economists and ecologists sometimes talk past each other. There is a feeling, at least among some ecologists, that economists ignore the vital role ecosystems play in supporting processes that sustain both human and ecological communities. There is a feeling among at least some economists that ecologists fail to appreciate the broad scope and versatility of economic methods in addressing concerns. These examples are presented with the intent of trying to overcome these mutual misunderstandings as well as further the dialog among ecologists and economists.

5.4.1 Statement of the Problem

To focus on important ideas that lead to potential disagreement, this first problem scenario is highly stylized, with details kept to a minimum. A business would like to build a commercial development on land that is currently forested. The project will generate positive economic returns to the business but will result in the destruction of a patch of forest habitat. An ecological survey finds a population of a rare and endemic species inhabits the forest patch. A decision maker, representing the public interest, will make a decision on whether to allow the development project to proceed. The decision maker is required by statute to consider both the economic returns to the business and the loss to society from any potential harm to the population.

Ecologist: Economists can't analyze a decision with long-run irreversible consequences like species extinction.
Economist: Oh yeah, we can do this.
Ecologist: Really? How?

5.4.2 Available Information

To keep the example quite simple, suppose there is a "current period" and a "future period" and that the discount factor applied to future period benefits and costs is 0.8 (corresponding to an interest rate of 25%). An economic study estimates that the project will generate returns of $20 million in the current period and $100 million in the future period. Based on current understanding of habitat requirements and population dynamics of the species, ecologists estimate that proceeding with the project will increase the probability of extinction from 10% to 30% in the future period. There is minimal risk of extinction in the current period regardless of whether the project is constructed or not. A contingent valuation survey of current residents of the area estimates that their willingness to pay to maintain the species is $500 million per period. Finally, suppose that once the forest patch is destroyed, its ecological value is lost and cannot be restored. In other words, there are irreversible consequences of the development.

Ecologist: While I have doubts about the accuracy of this information, I'll hold those questions. So here is your chance: show me how you would analyze the problem given this information.
Economist: Gladly. I am going to apply one of the standard tools of economics, benefit–cost analysis.

5.4.3 Application of Benefit–Cost Analysis

Benefit–cost analysis compares alternatives and requires that all relevant consequences of the alternatives be evaluated in the same metric, typically money. The analysis proceeds by calculating both the present value of the financial benefits of the development and the monetized costs of potential extinction. The present value of the financial returns of the project is (all values in millions of dollars)

$$B = 20 + (0.8)100 = 100$$

Complexity in Ecological Systems

The present value of the loss from increased risk of extinction is

$$C = 0.8(0.3 - 0.1)500 = 80$$

The present value of net benefits for the project is

$$B - C = 100 - 80 = 20$$

In this case, the benefits of the project exceed the costs, and the advice to the decision maker would be to proceed with the development.

Economist: See, this is a straightforward application of our standard set of tools.
Ecologist: But you have assumed that the *future value* of a species is based on *current preferences*. Surely that isn't the right approach.

5.4.4 Uncertain Future Losses from Extinction

The preferences of the future generation in the future period for the endemic species are unknown at present. A panel of experts is convened, and their conclusion is that there is a 50% chance that the next generation will be mesmerized by computer games and virtual entertainment and not care about this species, and a 50% chance that the next generation will be more concerned about species and have losses of $1 billion if the species is lost.

Benefit–cost analysis can still be applied, and in fact generates the same answer as before in terms of expected net benefits. The benefits are still $100 million. The expected costs are

$$C = 0.8(0.3 - 0.1) \times 0.5(1000) = 80,$$

which is exactly the same (expected) as before. The expected net benefits of the project are still $20 million.

With an irreversible decision and uncertainty, doing standard benefit–cost analysis as illustrated above is incorrect. An additional factor, "option value," must be factored into the analysis. Option value (referred to as "quasi-option value" in some literature) is the value of avoiding an irreversible decision prior to the resolution of uncertainty. The application of the option value to environmental or development issues was first developed by Arrow and Fisher (1974). Building the project in the current period is an irreversible decision. Once the project is built, the ecological value of the forest habitat is gone and the damage cannot be undone. But if the project is not built today, that does not prevent the future generation from deciding to build the project in the future.

Suppose the decision in the current period is to wait and let the future generation decide whether to pursue the project. Under the condition that the future generation places no value on the species, they will choose to build the project. This would generate benefits of $100 million and no costs. If the future generation places a value of $1 billion on species survival, then they will choose not to build the project. The expected net present value of waiting to decide is $(0.8)[(0.5)100 + (0.5)0] = 40$.

While the naïve benefit–cost approach would recommend that the project be built right away, a correctly specified economic analysis that factors in option value would recommend that decision makers avoid an irreversible decision until they learn whether the species is viewed as valuable or not by the future generation.

5.4.5 Additional Considerations

Economist: You are right that we shouldn't confuse *current* preferences with *future* preferences. But we have methods that take account of uncertainties and irreversibilities.

Ecologist: I see how it might be possible to take account of some uncertainty, but I still have a number of questions about your approach.

Preferences versus values: For some economists, these terms are synonymous. But there are other approaches that would not equate the 2. Some might say that values must include more than preferences, perhaps ethics, rights and responsibilities, fairness notions, and sustainability. An economist might counter that preference survey can be written in ways to include all these notions.... But the counter is that this does not really capture the other notions ... (and so on, back and forth, forever).

Ecological values: Often the focus of values generated from ecological systems is quite narrowly cast to include, for example, the increased value of commodity production, the value of an endangered species, carbon sequestration, or other "ecosystem services." Incorporating the full range of values generated from complex ecological systems is merely difficult at best and virtually impossible at worst. The value of ecosystem state and function may not be understood at present. How value would change under various stresses or anthropogenic change may be hard to estimate.

Pervasive uncertainty: We have discussed the difficulty in precisely valuing dynamic ecological systems. Equally difficult is determining trends in future preferences and human technology, so projected future costs and benefits have some appreciable associated uncertainty. Further, there are always questions about setting a discount rate, or whether to discount at all. In combination, these ecological and economic uncertainties may make the confidence interval of any results so large as to be useless for decision makers.

Risk versus uncertainty, or uncertainty versus ignorance: Typically, we really do not know the full extent of the event space, much less the probabilities associated with potential occurrences. Thus, a fair question is "Can probabilities be assigned to all future events?" Some argue yes, since Bayesian approaches involve assigning subjective prior probabilities based on whatever information is available. Others argue no, stating that since in complex systems some things are unknown, there will always be unanticipated, surprising events.

5.5 CASE STUDY: EXTRACTIVE DEVELOPMENT AND THE VALUATION OF BIODIVERSITY AND COMMUNITY NEEDS IN MADAGASCAR

5.5.1 Description of Complexities

Some of the following comparisons and analyses utilize information from the case studies at the end of this book.

The complexities of the system in Case Study 3, the development of a mineral extraction project in Madagascar, are significant if one considered only those associated with the natural system or, alternatively, the social-cultural system. However, to fully understand the issues and craft sustainable solutions, one must consider the ecological system as a whole, encompassing not only the flora and fauna endemic to the endangered littoral (coastal) forest but also the people who live and work near the forest. The people who live near the forest's, through engaging in traditional practices for gathering fuel and food, are using the forest's resources far faster than it can recover. Any solution to protect the littoral forest must also address the urgent socioeconomic needs of the people through practical, sustainable economic considerations that include capacity building at nearly all levels.

5.5.2 The Natural Environment

Madagascar is an isolated island nation (the 4th largest island in the world) located off the east coast of Africa. It is home to a large diversity of plant and animal species, many of them found nowhere else in the world. 85% of the plants and animals in Madagascar are unique to the island. Madagascar is regarded as one of the world's hotspots for biodiversity loss, as species disappear at an alarming rate in consequence to loss of the forests. Less than 10% of the original littoral forest remains on Madagascar, and much of what remains, present in small pockets along the length of the island, has been significantly degraded by human activity.

While understanding the structure and function of the littoral forest is essential to its protection and restoration, those understandings are not sufficient. The values of the local people with regard to the forest and its resources must be understood and their support obtained for any solution to be successful and sustainable.

The lack of established background data made comprehensive studies an essential part of planning from the outset. A significant program of baseline ecological and social research to evaluate the status of the forests and surrounding areas, and to begin understanding what values should be emphasized in restoration if the mining development was to be seen as sustainable, was embarked upon. A team of national and international specialists was commissioned in the late 1980s to undertake baseline studies of the existing natural and social environments. A conceptual mining project was outlined to define the area over which to carry out the studies. The study program was developed through discussion with a number of organizations, including the World Bank, the Canadian International Development Agency, Conservation International, the World Wide Fund for Nature in Madagascar, and other local interest groups. This work was carried out up to 1992 and concluded with a report available to all involved in the process.

5.5.3 PEOPLE WITH RELEVANT INTERESTS

A wide range of stakeholders have legitimate interests in the natural environment of southeastern Madagascar and the proposed mine development. Stakeholders in the project include the local Malagasy people, the local and regional governments, the national government, environmental and social NGOs, and intergovernmental aid agencies involved in the local area. These stakeholders have, in some cases, had conflicting values with regard to biodiversity, social-cultural traditions, economic opportunity, and poverty alleviation.

Local people value the forest as a source of wood for fuel, often in the form of charcoal, as well as a source of food. The lemur, several species of which live on Madagascar and are endangered, is a popular food for the local Malagasy. They use the forest to engage in a form of subsistence agriculture, *tavy*, which entails slashing and burning the forest. Roughly 70% of the population lives on less than $1 a day, below the World Bank's poverty standard. The local Malagasy, both in Fort Dauphin and in the smaller surrounding villages near the mining areas, currently are dependent on the shrinking forests to sustain themselves. And yet, there is a lack of understanding that there is a direct link between their traditional but destructive practices and the longer journeys required to find the forests and obtain fuel.

A Rio Tinto subsidiary, QIT Madagascar Minerals (QMM), has been conducting studies to evaluate mining sand deposits of ilmenite in southeastern Madagascar since the mid-1980s. The 3 large ilmenite deposits discovered by Rio Tinto in Madagascar cover roughly 6000 hectares and have an estimated worth of hundreds of millions of dollars in revenue. These deposits could provide up to 10% of world demand for ilmenite. The mine and the economic opportunity it will bring may offer the best chance of preserving some of the last remnants of littoral forest on Madagascar, by providing economic benefits and using the project to achieve broader biodiversity conservation goals.

Regional and national authorities recognize the linkages between the poverty of the citizens, the stewardship of natural resources, and the need for economic opportunity. The economy of Madagascar has been in decline for decades, and there is little external investment seeking entry. The proposed ilmenite mining project is regarded as an important component in the regional development of the southeast, and even if QMM decided against it, it is likely that extraction of the resource would continue regardless.

The project was visible to the international environmental and social nongovernmental and intergovernmental communities since its inception in the mid-1980s. Significant criticism was lodged against QMM and Rio Tinto for early development plans that specified mining through some of the last remnants of the littoral forest, but did not adequately articulate ecological mitigation alternatives or an integrated approach to the socioeconomic or cultural issues. Many of these early critics have since had the opportunity to examine the results of studies and consult on mitigation and management options being considered.

5.5.4 INTEGRATED SOLUTIONS

The baseline studies showed that conservation measures would be essential to protect species and suggested that a mining project could proceed in parallel with measures to preserve the area's most important environmental features. The studies proposed the creation of zones to conserve flora and fauna, including part of the littoral forest, while allowing other areas of the remaining forest to be gradually cleared over a period of many years. In these areas, it was suggested, replanting with a commercial forest crop of largely non-native species should follow any mining activity. These crops could provide an alternative source of fuel and construction wood for local people, thereby helping to relieve the present pressures on the forests.

Mined areas will be returned to natural contours, as has been done at similar mineral sand operations. Land immediately contiguous to the conservation areas will be rehabilitated in native species, while other disturbed land will be rehabilitated with fast-growing commercial species that will provide an alternative source of firewood, charcoal, and lumber for the local population. Much of the land in the mining area is already degraded by overuse. Highly valued wetlands, which are also a source of income to local people through the harvesting of reeds, will also be restored.

In order to address the challenges and commitments, QMN has taken a number of actions, including the following:

- Put into place a permanent social and environmental team of Malagasy professionals; there are more than 150 professionals and labor staff in the project area.
- Approached the management of the proposed conservation areas through dialogue led by the local authorities with the participation of all interested parties, leading to the adoption of a formal agreement.
- Actively supported a regional planning process, which is locally based and driven (and which has identified the mining project as a major positive factor in the development of the area).
- Planned and implemented a consultation process at the local, regional, national, and international levels with attention in preparing local consultations and adapting processes to local circumstances to facilitate maximum understanding of issues and values, while enhancing everyone's ability to express concerns and aspirations. Ongoing consultations help identify and test solutions.

5.5.5 SIGNIFICANT OUTCOMES AND LESSONS LEARNED

Lessons learned during the development of the project are described in more detail in Case Study 3. In brief, those lessons that most clearly speak to system complexity are as follows:

- Staff the environmental and social programs with full-time local employees.
- Integrate environment and social teams.

- Demonstrate to stakeholders that proposed restoration outcomes are feasible.
- Use outside experts to help address issues and solve problems.
- Complex problems cannot be fast-tracked.
- Flexibility in planning is required for credible stakeholder engagement.
- Support regional planning.

5.6 COMPLEXITY OF THE SYSTEM: CALFED CASE STUDY

A fundamental aspect of implementation of solutions to California's water issues, including the CALFED Plan (see Case Study 6), was the mandated designation of a volume of water to be used for environmental purposes. Resource management agencies were given the rights to use the water to support fish, other aquatic organisms, and the ecosystem supporting them in exchange for a greater assurance in maintaining exports for agriculture and urban consumption. There remains substantial uncertainty about whether the amount of water designated for environmental management is too much, too little, or just right. This is because of the complexity of determining how different biological attributes of the ecosystem respond to greater flows in time and space. Politically, there also seemed to be much more controversy when the currency was water than when the discussion centered on monetary exchanges. Perhaps this is because it is assumed that water used to support ecosystem functions is water lost to agricultural endeavors or other human benefit, while the money is just public money, not a direct loss to any specific individual.

One trade-off issue, widely debated, centered on the value of the water if it was used for agriculture versus the value of the fish saved by not exporting that water to the farmers. For example, in 2001, the number of winter-run salmon saved by curtailing pumping when fish migrate out through the delta was estimated as ~20000 salmon smolt. If 1000 outmigrating smolt died due to natural mortality for every returning adult produced, then saving those 20000 smolt amounts to saving about 20 adults. A typical curtailment might involve 50000 acre feet of water; at $10 per acre foot, these hypothetical numbers amount to a cost of $25000 per adult salmon saved. Stakeholders stop the discussion there and point to the intuitively outrageous cost per salmon. They then also ask for proof that saving 20 adults benefits winter-run salmon populations.

Calculations such as these obviously put pressure on ecosystem managers to more fully demonstrate the ecological benefits of curtailments. Some data exist, but the scale at which effects are detected is much broader than the scale at which the policy is being decided. For example, a relationship exists between river flows to the estuary and the abundance of some fish and invertebrate species important to the salmon food web. But those relationships occur over the entire range of inflows (natural inflows vary from 5 million acre feet [MAF] in the driest years to 35 MAF in the wettest years). The scale is too coarse to allow evaluations of changes as small as 1 MAF, except at the very lowest flows.

In mark-recapture experiments, statistical indications are that the fewest salmon survive when traversing the delta during years of lowest flows and highest pumping rates. But again, the grain of the data is too coarse to relate specific numbers of salmon lost (or any other species) to changes as small as 1 MAF.

It is also conceivable that pumping these massive quantities of water could affect the physical system, the fish movement, and the food web in ways difficult to translate into numbers of adult fish. But, again, existing methodologies are too insensitive to understand even the potential effects of such changes. Indirect benefits, like increased population resilience, may occur when higher flows allow some individuals to spend more time in the delta nursing prior to heading out to the ocean. This effect has been shown in other systems, but such benefits remain undocumented in the delta, and thus remain controversial.

There is a mismatch in this circumstance between knowledge of benefits of environmental water for salmon in the complex delta ecosystem and knowledge of potential benefits to agriculture of using the same water. Salmonids have commercial benefit, but it is difficult to value that because any increase in the population in response to ecosystem protection measures is difficult to determine. And it is not clear whether the limits to the agencies' ability to prove benefits are because those benefits are small or because of type II error in detecting the benefits (insufficiently sensitive analytical tools). One outcome of the disparities in complexity may also be a subtle shift of the burden of proving beneficial effects toward managers dealing with the more complex fish issues. Ultimately, the shift in burden of proof could affect political decisions about the trade-offs.

Avoiding extinction is a well-recognized value and a critical point that economists recognize as beyond valuation. But the threat of extinction is not irreversible. In fact, some species have been delisted under the ESA when key threats were removed (delisting of the brown pelican after the banning of organochlorine compounds and the recovery of populations is an example). Delisting raises important questions about predicting long-term welfare of a species. The complexity of these issues is great, and the answers are not especially well known in ecology.

While declines in winter-run Chinook salmon were strongly evident from 1970 through the early 1990s, abundances have increased since the mid-1990s. This increase in abundance coincides with actions to engineer new cold water habitat, build hatcheries for this specific run, invest in habitat restoration efforts, and reduce harvesting pressures in the ocean. The coincidence with these actions makes it tempting to conclude that the investments in winter-run salmon have this animal on the way to recovery.

But ecologists have all called for caution in interpreting these numbers. Natural cycles of abundance are well-known for all kinds of animal populations. The linkage of salmon abundances to a 20-year cycle in ocean conditions (the Pacific Decadal Oscillation) was well demonstrated in the Pacific Northwest. Its application to California waters is less well studied but nevertheless likely. Several studies have shown that a salmon population under constant anthropogenic stress will decline in progressive steps rather than linearly. During good ocean conditions, populations may increase relative to previous lowest levels; but when ocean conditions turn bad, the accumulation of the effects of human disturbance is even more strongly expressed in new lows in abundance. Such patterns are conceivable in the winter-run data.

But the period of record is short relative to such cycles (even though it is 30 years long). So it is not yet clear that the recovery reflects partial response to natural conditions (to be followed by an even stronger collapse) or a true recovery. Human stresses have been reduced, but only more time will tell if the reduction is adequate to declare the species safe from extinction.

5.7 PCB CASE STUDY

Most decisions that involve complex ecosystems will have a series of management objectives corresponding to particular regulatory requirements and frameworks. It is not uncommon to find that the management goals and objectives established under one regulatory environment generate particular kinds of data and analyses that are suitable for that specific evaluation, but may have limited utility when applied in another regulatory context. For example, there is often a succession of ecosystem studies conducted under USEPA regulatory authority, such as the Comprehensive Environmental Response, Compensation, and Liability Act (CERCLA) or Superfund, to then be followed by a natural resource damage assessment (NRDA) under the regulatory guidance of the US Department of the Interior (DOI). Each regulatory framework has its own objectives and goals, some of which are overlapping and some of which are distinct. The goal of a typical Superfund assessment is to determine the potential for significant risk or hazard as a result of exposure to contaminants, and to evaluate the efficacy of potential remedial alternatives (including no action or some type of monitored natural attenuation). The goal of an NRDA, by contrast, is to establish injury for the purpose of restoration. Although these 2 objectives have methods and approaches in common, in practice, there is often little communication between the 2 regulatory programs.

There are numerous PCB-contaminated rivers in the United States, including the Fox River, Sheboygan River, Grasse River, and Hudson River. In some cases, these sites are large and complicated enough to warrant listing as Superfund sites. In addition, there are often extensive repercussions from potential exposures leading to the requirement for an NRDA. Establishing injury typically requires field data on changes in key endpoints relevant to sustaining populations. Similarly, determining risk can benefit from these kinds of studies, and in fact, the determination of potential or no potential significant risk is supported by these same kinds of studies.

Another frustration in relating analyses supporting the risk assessment to a quantitative determination of injury is that typical ecological risk assessments do not express risk as a probability. Following USEPA guidance (USEPA 1997, 1998), calculated exposure doses for the receptors of concern are compared to toxicity reference values. The resulting ratio has no meaningful interpretation. Is a toxicity quotient of 100 ten times worse than a toxicity quotient of 10? And what is the probability of exceeding a particular threshold?

Probabilistic approaches, particularly for exposure but even toxicity in the case of ecological risk, are gaining favor (USEPA 2001). These more sophisticated approaches allow analysts to quantify the probability of an increasing magnitude of effect. This is a welcome change in methodology that will facilitate integration with economic analyses.

Complexity in Ecological Systems

5.8 LESSONS LEARNED AND NOT LEARNED

> In anything at all, perfection is finally attained not when there is no longer anything to add, but when there is no longer anything to take away.
>
> **Antoine de Saint-Exupéry,** *Wind, Sand and Stars* (1940)

5.8.1 INTEGRATING ECOLOGY WITH ECONOMICS

Opportunities exist for effective integration of ecology and economics based on an evaluation of scientific study outputs (data or modeling results) serving as inputs to economic valuation studies. For example, a probability of a particular ecological effect related to population sustainability can be directly valued through the use of survey methods.

Further, the concept of ecosystem services offers a common point of reference for dialog between economists and ecologists. There are tools and approaches appropriate for use in evaluating the economic implications of changes to ecosystem processes resulting from management actions (including taking no action). Ecological resource valuation can use nonmarket-based economic assessment tools, with assessed values reported in nonmonetary (or monetary) terms. Finally, sequential data handoffs and concurrent or parallel analyses offer much less effective guidance to management and little potential for creative outcomes. The concept that integrating economic and ecological approaches early in the assessment process can yield unexpected, positive consequences presents the best lesson learned.

5.8.2 APPRECIATING AND UNDERSTANDING COMPLEXITY

An appreciable lack of understanding of complex ecological systems can preclude useful integration with economics. Having a clearly defined conceptual model — in terms of both system complexity as well as how management goals may alter system behavior — leads to the effective combination of tools and techniques across disciplines. A conceptual model must consider spatial and temporal scale when putting into context regulatory and ecological regimes. A conceptual model provides the foundation for any analyses used to quantify ecosystem processes. The complexity captured in a conceptual model needs to reflect the complexity of the system at the chosen level of analysis. It also needs to adequately capture system responses to management actions. In addition, maintaining similar levels of analysis between the ecological and economic aspects of an issue helps to maintain a consistent and integrated conceptual model. A conceptual model that captures relationships and processes at the level of observation and analysis is essential to defining measurement and assessment endpoints (both economic and ecological) at appropriate spatial and temporal scales. It is also essential to understand how the conceptual model at this level relates to the overall ecosystem structure at broader spatial and temporal scales.

Though conceptually different, both uncertainty and variability represent required considerations in any model or analysis. Even with the uncertainty present in every quantitative and predictive model, bounding ecological uncertainty greatly increases the usefulness of any economic valuation. This way, economists can take advantage of uncertainty as readily as absolute numbers. Alternative approaches might include a best estimate of associated uncertainty, or use forecasts built from scenarios. As well, decision makers need not accept "bright line" or single-number analyses of risk or cost. Such analyses provide a false sense of confidence in the result. For example, placing a value on extinction, an irreversible state, presents a near impossibility. Yet, there still remains great uncertainty concerning the actual level of the threat of extinction. This uncertainty provides an opportunity to use trade-off analysis tools to understand management options. Decision analytic frameworks exist for conducting trade-off analyses (see Chapters 3 and 4). These tools provide a means for integrating ecological and socioeconomic data and model results, and provide transparency to the decision-making process. These kinds of decision support tools help to integrate the results of ecological analyses with stakeholder preferences, costs, and other socioeconomic considerations.

Rather than seeking to achieve a predetermined state in a complex system, identifying an appropriate change in trajectory may represent a better indicator of success, as well as evaluating the rate of system inputs and outputs over monitoring absolute numbers or changes in numbers of things (e.g., biodiversity) inside an ecosystem. Every local analysis must consider the broader context. In ecological analyses, there often exists a bias in focus toward "returning the system to the way it was" or some other predetermined state. However, given the constant flux of ecosystems, it makes little sense to establish a static parameter set as a realistic management goal. Much like environmental parameters, societal values also change over time. Management approaches and goals need to recognize the dynamic nature of that which is valued as well as the values themselves. The fact is that there is no single scientifically based approach that is preferable to another. "Preferable" is the result of values, not science.

Useful valuation does not require monetization. Managers can make value-based decisions that protect the public interest without needing monetization as the economic tool. Money (e.g., US dollars, yen, and euros) represents only one of the many different ways to express value. Using decision-making processes that lean too heavily on traditional currencies in complex systems can greatly undervalue ecosystem services. Most importantly, make sure the burden of proof does not shift, consciously or unconsciously, to those responsible for managing the most complex resources, which typically represent the most difficult valuation situations.

New knowledge of system dynamics can open alternative pathways that lead to new management approaches. Both the environment itself and our understanding of it dynamically and constantly evolve. Consequently, the perceived, present-day value of any ecosystem attribute can change dramatically with new knowledge.

Beneficial and long-term solutions structure value trade-offs so as not to appear a zero-sum game. There always exists a tension between near-term and local socioeconomic needs versus long-term and global ecological values, as revealed by the Madagascar case study. Other examples, such as the atmospheric transport and deposition of PCBs or mercury, show how local outputs can have global implications, or vice versa. Finally, any valuation of a complex ecological resource needs to consider positive or negative changes across appropriate spatial and temporal scales.

REFERENCES

Ahl V, Allen TFH. 1996. Hierarchy theory: a vision, vocabulary and epistemology. New York (NY): Columbia University Press.
Allen, TFH. 1998. The landscape "level" is dead: persuading the family to take it off the respirator." In: Ecological scale: theory and applications. Peterson, David L & V. Thomas Parker, editors. New York (NY): Columbia University Press. p 35–54, Chapter 4.
Allen TFH, Hoekstra TW. 1992. Toward a unified ecology. In: Allen TFH, Roberts DW, editors. Complexity in ecological system services. New York (NY): Columbia University Press. 384 p.
Allen TFH, O'Neill RV, Hoekstra TW. 1987. Interlevel relations in ecological research and management: some working principles from hierarchy theory. J Appl Sys Anal 14:63–79.
Arrow K, Fisher AC. 1974. Environmental preservation, uncertainty, and irreversibility. Quart J Econ 88:312–319.
Baveye P, Boast CW. 1999. Physical scales and spatial predictability of transport processes in the environment. In: Corwin DL, Loague K, Ellsworth TR, editors, Assessment of non-point source pollution in the vadose zone. Geophysical Monograph 108. Washington (DC): American Geophysical Union, 261–280.
Bengtsson BE. 1980. Long-term effects of PCB (Clophen A50) on growth, reproduction, and swimming performance in the minnow, *Phonixus phonixus*. Water Res 14:681–687.
Brown, CL, Parchaso, F, Thompson, JK, and Luoma, SN 2003. Assessing toxicant effects in a complex estuary: a case study of effects of silver on reproduction in the bivalve, Potamocorbula amurensis, in San Francisco Bay. The International Journal of Human and Ecological Risk Assessment 9:96–119.
Cairns, J, Jr. 1990. The threshold problem in ecotoxicology. Ecotoxicology 1:3–16.
Cairns, J, Jr. Mount, DI. (1990). Aquatic toxicology. Environmental Science and Technology 24:154–161.
Cloern JE, 2001. Our evolving conceptual model of coastal eutrophication. Mar. Ecol. Prog. Series 210:223–253.
Hornberger MI, Luoma SN, Cain D, Parchaso F, Brown CL, Bouse RM, Wellise C, Thompson JK. 2000. Linkage of bioaccumulation and biological effects to changes in pollutant loads in South San Francisco Bay. Environ Sci Technol 34:2401–2409.
Kay JJ, Regier H, Boyle M, Francis G. 1999. An ecosystem approach for sustainability: addressing the challenge of complexity. Futures 31(7):721–742.
Leopold, A. 1948. A Sand County almanac, and sketches here and there. New York (NY): Oxford University Press, 256 p.
Long ER, Field LJ, MacDonald DD. 1998. Predicting toxicity in marine sediments with numerical sediment quality guidelines. Environ Toxicol and Chem 17:714–727.
Luoma SN. 1996. The developing framework of marine ecotoxicology: pollutants as a variable in marine ecosystems? J Exp Mar Biol Ecol 200:29–55.
Luoma SN, Clements WH, DeWitt T, Gerritsen J, Hatch A, Jepson P, Reynoldson T, Thom RM. 2001. Role of environmental variability in evaluating stressor effects. In: Baird DJ, Burton GA Jr., editors, Ecological variability: separating natural from anthropogenic causes of ecosystem impairment. Pensacola (FL): SETAC Press. 307 p.
Morgan MG, Henrion M. 1990. Uncertainty: a guide to dealing with quantitative risk and policy analysis. New York (NY): Cambridge University Press.
Presser TS, Luoma SN. 2006. Forecasting selenium discharges to the San Francisco Bay–Delta estuary: ecological effects of a proposed San Luis drain extension. US Geological Survey Professional Paper 1646. 196 p. http://pubs.usgs.gov/of/2000/of00-416/pdf/OFR-00-416.pdf.
Rittel H, Webber M. 1973. Dilemmas in a general theory of planning. Pol Sci 4:155–169.
Saint-Exupéry, A de. 1940. Wind, sand and stars. New York (NY): Harbrace.

Thompson KM, Graham JD. 1996. Risk assessment to improve risk management. Human Ecol Risk Assess 2:1008–1034.

[USEPA] United States Environmental Protection Agency. 1997. Ecological risk assessment guidance for Superfund: process for designing and conducting ecological risk assessments. Interim final. Edison (NJ): USEPA, Environmental Response Team. http://www.epa.gov/superfund/programs/risk/tooltrad.htm#gdec.

[USEPA] United States Environmental Protection Agency. 1998. Guidelines for ecological risk assessment. USEPA EPA/630/R095/002F, April 1. Washington (DC): USEPA, Risk Assessment Forum. http://cfpub.epa.gov/ncea/cfm/recordisplay.cfm?deid=12460.

[USEPA] United States Environmental Protection Agency. 2001. Risk assessment guidance for Superfund: Volume III — part A, process for conducting probabilistic risk assessment. EPA 540-R-02-002, December. Washington (DC): USEPA.

Von Stackelberg K, Vorhees D, Linkov I, Burmistrov D, Bridges T. 2002. Importance of uncertainty and variability to predicted risks from trophic transfer of contaminants in dredged sediments. Risk Anal 22(3):499–512.

Yoe CD. 2002. Trade-off analysis planning and procedures guidebook. IWR-02-R-2. Washington (DC): US Army Corps of Engineers, Institute for Water Resources. http://www.wrc.usace.army.mil/iwr.

6 Organizing and Integrating the Valuation Process

Brian Heninger, Greg Biddinger, Chester Joy, Catherine L. Kling, Doug Reagan, and Travis S. Schmidt

CONTENTS

6.1	Introduction	130
	6.1.1 Value-Based Decision Making	131
	6.1.2 Principles and Characteristics of an Effective Evaluation Process	131
	6.1.3 Chapter Overview	132
6.2	Framework Comparisons	132
	6.2.1 Review of Existing Regulatory Frameworks	132
	6.2.2 Review of Existing Environmental Management Frameworks	134
	6.2.3 Synthesis	135
6.3	Influences of the Management Context on Integrating the Valuation Process Into Environmental Decision Making	136
	6.3.1 Elements of Context	136
	6.3.2 Theoretical and Conceptual Valuation Process and the Role of Public versus Private Decision Making	136
	6.3.2.1 Private Firms	136
	6.3.2.2 Local Communities	137
	6.3.2.3 Public Decision Makers	138
	6.3.3 Management Context	138
	6.3.3.1 Decision Purpose	138
	6.3.3.2 Decision Scope	139
	6.3.3.3 Decision Authority	141
	6.3.4 Management Context Influences on Integration: Current Limitations and Opportunities for Improvements	143
	6.3.4.1 Current Limitations on Integration within the Management Context	143
	6.3.4.2 Identifying Opportunities for Better Integration	145

6.4 Comprehensive and Systematic Approach to Identifying
      Assessment Endpoints 146
      6.4.1   Assessment Endpoints 146
      6.4.2   Problem Identification and Goal Setting 147
      6.4.3   Stakeholder Identification 148
      6.4.4   Identification of Assessment Endpoints 149
              6.4.4.1   Identifying Relevant Ecological Properties 150
              6.4.4.2   Identifying the Human Values of Ecosystem
                        Components 154
      6.4.5   Data Collection 156
      6.4.6   Using Ecology and Economics to Identify Endpoints for
              Valuation and Provide Information for Decision Making 157
6.5 Informing the Valuation Process: Summary 160
References 161

6.1 INTRODUCTION

The focus of this chapter is on understanding and meeting the information needs of the environmental decision maker for different types of decisions. Specifically, in the area of valuation, where analysts coming from disparate backgrounds can provide alternate and sometimes conflicting information, it is critical to present the results of evaluations in a comprehensive, systematic, and understandable framework so that decision makers can consider the full range of relevant information.

The previous chapters have focused on describing a variety of methodologies for valuation, their conceptual underpinnings and complexities associated with various methods. As we have seen, there are many approaches to the valuation of ecological resources, from economic valuation, social and cultural valuation, and many other methods for assigning quantitative information to an ecological endpoint. The current chapter adds perspective to the previous discussions of valuation by addressing the issues of decision context, the role of valuation in the decision process, and how integration within the valuation and decision process can lead to more informed decisions.

All environmental decisions are made on the basis of values, and all values are ultimately human based (i.e., the value that humans place on an object, process, or outcome). Even ecological attributes are human-based values because they are derived from humans' understanding of how ecosystems work. Humans need to understand how an ecosystem produces the goods, services, and characteristics that we value. This understanding also changes over time, for example, 50 years ago wildfires were considered "bad," but recent studies have shown the benefit of fire to ecosystem structure and function, thus enhancing its ability to provide the goods and services that humans value.

Ecologically oriented valuation approaches have generally focused on the various levels of ecological organization from ecosystem, community, species, population, and individual. Early economic-based valuation approaches that attempted to place a monetary value on ecological properties experienced limited success. Recent methodologies that focus on services provided by ecosystem components, to either the ecosystem or the human uses thereof, have provided useful tools to link ecological attributes with economic valuation.

6.1.1 Value-Based Decision Making

Environmental management decision processes are essentially driven by the values associated with the decision at hand and the context in which it will be made. Information to support these decisions can come from a range of disciplines in the environmental (ecological risk assessment, environmental impact assessments) and social sciences (benefit–cost analysis and social and cultural impact assessments). Since practitioners in these varied disciplines support management's ability to make effective decisions with different tools, early and frequent coordination among the assessors would provide greater assurance of an integrated assessment. Therefore, ecological, economic, and social-cultural assessors need to be involved in the planning process before the assessments are initiated. Early identification of critical ecological endpoints is fundamental to the process.

The lack of systematic integration of environmental and social sciences has led to complaints from some economists that they are not brought into the assessment process early enough to properly design their analysis. Joint planning by the participating assessors would allow the economists to understand what is important to the environmental scientists in the way of ecological protection goals. Similarly, it would provide the environmental assessors with grounding in what ecological services or functions can be assessed with economic tools. Thus, if there is any flexibility in selection of the environmental assessment or measurement endpoints, then the ability to subsequently perform aligned assessments will be enhanced.

Ultimately, all of the assessments need to be designed to optimize their service to the manager in making a decision. The most effective decisions will be generated if all of the essential disciplines are engaged throughout the process from the planning of the assessment to communication of the results to inform the actual management decision. In a perfect world, any environmental assessment (e.g., ecological risk assessment) will address the ecological services and functions associated with the natural resources valued by those affected by the outcome of the decision.

Value-based decision making recognizes that all decisions are ultimately made on the basis of what people value, but many such decisions have not involved all relevant stakeholders in the decision-making process. In ecological risk assessments, ecological attributes are often listed in terms of selected ecosystem properties or components such as biodiversity. Recent assessments have successfully implemented an approach that identifies the ecological characteristics to be protected, sustained, or restored using a systematic and comprehensive approach for identifying all possible attributes as a basis for impact assessment or environmental management planning (see section 6.4). This approach integrates ecological, social, and economic concerns into the evaluation process.

6.1.2 Principles and Characteristics of an Effective Evaluation Process

Both environmental and economic assessments need to reflect the attributes of effective and high-quality environmental management. Such principles or characteristics of effective management processes have been addressed in international environmental management standards (ISO 2004). These include considerations of transparency, participation, accountability, and adaptability.

6.1.3 CHAPTER OVERVIEW

This chapter describes the ecological and economic information sets, tools, and analytical frameworks that are needed to make and support sound environmental decisions, and discusses how these may vary according to the type of decision being made. The following 4 sections include discussions and comparisons of the frameworks that exist to organize and guide environmental decision making, and how they have evolved over time, followed by a discussion of the influence of management context on identification of assessment endpoints and the information necessary for decision making. Then a description of the process of endpoint identification is presented for both endpoints, which are valued directly, and those that are valued indirectly (i.e., because of their contribution to the production of ecological endpoints that are valued directly). Lastly, a section on the integration and communication of ecological assessment and valuation is presented. The theoretical or technical adequacy of any given decision-making or valuation approach will not be undertaken in this forum (but see Chapter 4). Throughout this chapter, the emphasis will be on how integration between disciplines can improve the process of producing and presenting information that can better aid the decision maker in making well-reasoned, balanced decisions in a variety of settings.

6.2 FRAMEWORK COMPARISONS

Environmental management decision making has long been recognized as an integrative and complex task. A necessary process to making complex decisions is the establishment of a system by which data are collected, processed, evaluated, and then integrated with other types of information. In virtually all ecological risk assessment guidance documents, a conceptual paradigm, often called a "framework," is offered to help establish such a system. The initial frameworks were constructed to guide the collection of data for generating scientifically defensible information for decision makers about the adverse effects and risks due to chemical exposure to humans. Over time, society demanded the expansion of the risk paradigm to include the ability to assess nonchemical risks and risks to ecological systems. As the paradigm for ecological risk assessment (ERA) expanded, so did the need to incorporate economic and social tools to quantify risks. The following narrative will not try to expand the large number of frameworks already available to risk assessors, but will try to synthesize some of the history of the integration of environmental, economic, and social science into the ecological risk assessment paradigm.

6.2.1 REVIEW OF EXISTING REGULATORY FRAMEWORKS

In the early 1980s, the US government began producing legislation mandating the management of chemical risks to humans and the environment (i.e., the 1980 Comprehensive Environmental Response, Compensation, and Liability Act and the

1986 Superfund Amendments and Reauthorization Act). In response, the National Research Counsel (NRC) produced a document that developed the initial paradigm for the assessment of the risks to human health resultant from exposure to chemicals (NRC 1983). This paradigm has 4 phases:

- Hazard identification
- Dose–response assessment
- Exposure assessment
- Risk characterization

The NRC paradigm was the first risk assessment paradigm to be incorporated into guidance documents in the US Environmental Protection Agency (USEPA) Office of Solid Waste and Emergency Response (OSWER). From this paradigm, frameworks for human health risk assessments and ecological risk assessments were developed (USEPA 1989a).

Although guidance for ecological risk assessment was developed for remedial investigations, the USEPA continued to refine guidance for ecological risk assessment. A series of advisory boards was developed to investigate the appropriate paradigm for use in ecological assessment (NRC 1983). The USEPA's Risk Assessment Forum concluded that the NRC paradigm would require modification for application in ecological risk assessments. Following the Risk Assessment Forum, the Science Advisory Board (SAB) advised the USEPA to broaden its scope on ecological risk assessment from its exclusive focus on chemical risks to incorporating nonchemical risks as well (USEPA 1990). The "Framework for Ecological Risk Assessment" (USEPA 1992; hereafter referred to as the Framework) was the result of these efforts.

This Framework was the first guidance for ecological risk assessment intended for use outside of the Superfund Program. The paradigm in the Framework defined the role of the ecological risk assessor as being limited to performing the risk assessment on a technical basis, while the risk manager communicated society's values when appropriate and prescribed by policy or legal mandate. The Framework outlined a 3-step iterative process:

- Problem formulation
- Analysis
- Risk characterization

This new paradigm has been modeled throughout other documents providing guidance on performing ecological risk assessments since 1992. Other distinguishing characteristics of the Framework that are not found in the human health risk assessment paradigm are the consideration of effects beyond individuals, to include populations, communities, and/or ecosystems. Problem formulation is an initial step to develop the focus and scope of an ERA, and the term "stressor response" is utilized rather then "dose response," providing the latitude to consider nonchemical stressors as well as chemical stressors (e.g., invasive species, biocontrol, and genetically modified organisms). Other additional novelties found in this document included the incorporation of the notion of "environmental values" and the need for a multidisciplinary team as a prerequisite of a successful ERA.

While the Risk Assessment Forum was refining the ecological risk assessment paradigm and generalizing ERA to include nonchemical stressors, the USEPA's OSWER was also refining the guidance for ecological risk assessment for Superfund (USEPA 1997). This document was developed by OSWER to accompany the Framework. This document moved ERA forward in the Superfund program, adopting the ERA paradigm created by the Risk Assessment Forum, and expanding the paradigm to an 8-step iterative process with an increasingly narrow focus of investigation. More importantly, this new guidance document, for the first time, explicitly included the incorporation of input from parties of interested stakeholders that were affected by the object of concern in the problem formulation phase.

The explicit inclusion of stakeholders and incorporation of local social-economic factors in the 1997 OSWER guidance document was a significant departure from the paradigm designed exclusively for generating scientifically defensible information for decision makers. The Framework relied on risk managers to consider societal values and policies as defined in USEPA statutes.

In the USEPA's "Guidelines for Ecological Risk Assessment" (1998), more explicit social and economic factors were incorporated into the ERA paradigm. Based on the Framework, this expanded paradigm explicitly defined the roles of the risk manager, risk assessor, and other interested parties in the problem formulation phase of ERA. It is also suggested that the expanded paradigm provides a critical element for environmental decision making by giving risk managers an approach for considering available scientific information along with the other factors they need to consider (e.g., social, legal, political, or economic factors). In this paradigm, social and economic factors are given an expanded role. Previously, these factors were implicit in the risk manager's decision processes.

The expansion of the risk paradigm and the broadened mandate by which ERA is to be conducted have created momentum in efforts to incorporate social and economic tools into ERA. Over the years, the USEPA has formed a number of workgroups and committees to address these issues, and has produced guidance to stimulate its incorporation. The "Guidelines for Preparing Economic Analyses" was published by the USEPA's National Center for Environmental Economics in 2000 (USEPA 2000b), and the USEPA published "A Framework for the Economic Assessment of Ecological Benefits" in 2002 (USEPA 2002). In addition, the USEPA's National Center for Environmental Assessment has communicated an approach specifically integrating ERA and economic analysis in "Integrating Ecological Risk Assessment and Economic Analysis in Watersheds: A Conceptual Approach and Three Case Studies" (USEPA 2003).

6.2.2 Review of Existing Environmental Management Frameworks

Along with the development of frameworks for risk assessment, there has been a number of management paradigms developed to help guide decision makers (PCCRARM 1997a, 1997b; Stahl et al. 2001). Many of the management paradigms are based on the paradigm utilized in the USEPA (1992) Framework and USEPA

(1998) Guidelines. The Presidential/Congressional Commission on Risk Management developed guidelines for management decision making and risk assessment guidelines to help ensure good risk management decisions emerged from a decision-making process that elicits the views of those affected by the decision, so that differing technical assessment, public values, knowledge, and perception were considered (PCCRARM 1997a, 1997b). It was also the goal of these guidelines to synthesize the roles of environmental and public health agencies through a paradigm by which each stage relies on broader contexts, and stakeholder participation, in an iterative process.

The USEPA Science Advisory Board has also developed paradigms for managers and decision makers (USEPA 2000c). The SAB intended to create the "next step in the involvement of a wider range of people and perspectives/values in the decision making processes." The SAB also acknowledged that fragmentation in the environmental decision-making process had resulted from the statutes that address environmental problems chemical by chemical, or medium by medium. Individual statutes fragment the process by preventing the co-management of multiple stressors in the environment and preventing the establishment of criteria an agency might use to set priorities that cut across statutory boundaries (NAPA 1995). The SAB encouraged integration of ecological, economic, and social tools to evaluate integrative data together and not by data type.

The majority of the ecological risk management (ERM) paradigms written for the audience of decision makers and risk managers are based on the USEPA (1992) Framework and USEPA (1998) Guideline documents. Most of these documents make it explicit that risk assessors perform the assessment, while risk managers and decision makers are the only people officially involved in the risk assessment process who consider social or economic factors in environmental management decisions. With the continued push to involve social scientists, economists, and stakeholders at the initial steps of risk assessment in most contemporary ERA paradigms along with the environmental risk assessors and managers, the clear line between scientific process and social-economic process is fading. Ecological risk assessors are now asked to be scientifically objective while performing risk assessment in an atmosphere of social and economic awareness never expected before.

6.2.3 SYNTHESIS

Environmental risk assessment began as the analysis of chemical exposure to humans and the potential for adverse effects. Paradigms were developed to guide a generation of scientifically defensible risk assessments. These paradigms considered social values and the economic impact of such risk assessments; however, by what means was not made explicit. As the mandate for environmental risk assessment broadened to include nonchemical risks as well, it became increasingly clear that the contribution of scientists from different disciplines would be required for the successful completion of the risk assessment. Current paradigms for environmental risk assessment and decision making explicitly incorporate economic and other social factors.

6.3 INFLUENCES OF THE MANAGEMENT CONTEXT ON INTEGRATING THE VALUATION PROCESS INTO ENVIRONMENTAL DECISION MAKING

6.3.1 ELEMENTS OF CONTEXT

When discussing decision making, especially from the perspective of valuation, it is important to first identify all of the elements of the decision context. The valuation process will differ based on the context of the decision to be informed. Management contexts include decision purpose or type (e.g., public or private), decision scope (scale, focus, and goals), and decision authority (environmental laws and general management statutes). The determination of the decision context will help establish the framework for the decision process and have critical implications for valuation as well.

6.3.2 THEORETICAL AND CONCEPTUAL VALUATION PROCESS AND THE ROLE OF PUBLIC VERSUS PRIVATE DECISION MAKING

In considering the role that valuation of ecological resources (explicitly quantified or not) can or should have in a decision, it is critical to first understand the circumstance of the decision maker. A key part of this context is whether the decision maker is acting on behalf of an individual firm, a community, or a larger social group such as a government entity, nongovernmental organization (NGO), or conservation agency. In general, different concepts and therefore measures of value will be relevant to each level of decision making and different types of decisions. We illustrate this point by discussing some of the situations in which different kinds of decisions could be aided by valuation of ecological resources.

6.3.2.1 Private Firms

In the case of a decision maker acting entirely on behalf of a private enterprise, it is reasonable to think that he or she will be interested in the value of an ecological resource only in so far as he or she can capture some or all of that value directly or indirectly to benefit the firm. Direct benefits would occur if the firm could increase revenues from its products or decrease the costs of producing its products by protecting the ecological resource, thereby raising profits. Examples of indirect benefits that could be obtained by the firm that acts on behalf of the environment are improved relationships with its local community, improved firm image (possibly resulting in increased sales), and better working conditions for its employees.

In the economics literature, increased product prices for firms that produce and sell ecologically friendly products are termed a voluntary provision of public goods. Numerous examples of studies from economics and the marketing literature have explored these price effects (e.g., Videras and Alberini 2000; Khanna 2001). Producers may try to appeal to consumers who demand "green products," or preempt government regulation. Quantification of the indirect benefits in terms of social goodwill may be harder to quantify, but are well understood by firms, and are routinely included in the calculation of the worth of a firm's brand name.

Organizing and Integrating the Valuation Process

While the general concepts of benefit–cost analysis are relevant in the case of private decision making, these private decision makers can be expected to respond only to the benefits and costs that affect their individual enterprise and not the benefits or costs that accrue elsewhere. In a private business context, a firm will only take an action (including environmental control) if the benefits (to itself) of doing so outweigh the costs. Typically, fines and penalties have been used to induce firms to make environmentally friendly decisions. However, as noted above, there is a vast amount of literature devoted to what motivates firms to voluntarily reduce emissions or protect an ecologically important resource. If one dismisses pure altruism as a motivation, then the firm must be getting something back in return from their investment in pollution control. Whether it is a better standing as a good environmental citizen or the decreased probability of regulation, their actions are valued as a benefit to themselves, and the firm must estimate that monetary benefit to themselves.

6.3.2.2 Local Communities

Another type of decision maker who might be interested in values concerning ecological resources is a business owner who does not directly produce or own an ecological resource, but whose business is affected by the presence or absence of such a resource. More generally, area chambers of commerce may represent the interests of numerous such business owners and citizen groups. For example, a small community may be located near a lake predominately used for recreation. Changes in the water quality of that lake may affect the business activity in that town and therefore tax collections for local coffers and the ability for local citizens to earn their living nearby.

In such a circumstance, the community may be interested in the gains that they can garner from the resource (the lake and, more specifically, the level of its water quality). For example, suppose a project that would clean up the local lake is being considered. The local residents and community leaders might be interested in the amount of new business this might bring to the town. These impacts can be cumulative since when business increases at a local restaurant, part of that money will be respent in the local area at other businesses and on wages. Thus, there is a "multiplier" effect of business activity.

Economists and regional planners refer to measures of this sort as the economic impact of a resource, and there is a large literature termed "economic impact analysis." Numerous textbooks on the topic identify the appropriate methodology for computing multiplier effects and other related concepts.

From a valuation perspective, most economists would agree that these "economic impacts" do not measure the value of the improvement in water quality. Rather, they are pecuniary effects associated with its improvement that accrue to the local citizens. However, these local impacts are largely transfers of economic activity from one location to another. This is because if the water quality of the lake were not improved, the meals that would have been eaten at the local restaurants would be eaten elsewhere; that experience does not disappear, but is just placed elsewhere. Thus, while enjoyment of the meal may be higher given its location, the expenditures for the meal do not capture the value of the water quality improvement. Economic impact measures do not capture the value of the resource per se; they do, however, measure the gains to local businesses and are thus of real interest to some.

6.3.2.3 Public Decision Makers

Valuation information used by a decision maker representing a government organization is likely to be very different from either of the previous cases above. In this situation, the full social benefits and costs of a project are incorporated in any benefit–cost analysis.

The fundamental goal of a benefit–cost analysis is to provide decision makers with an aggregate estimate of the economic benefits and costs of a project or change. Roughly speaking, the economic benefits incorporate the total willingness to pay by all affected individuals to have the project or change. Similarly, costs are the total willingness to pay to avoid the change or the costs associated with the project. As such, economic valuation methods conceptually incorporate broadly the trade-offs people are willing to make to preserve environmental resources and provide a framework to compare the costs and benefits.

In addition, a thorough benefit–cost analysis will provide information on the distributional consequences of the proposed change (i.e., which groups benefit from the change, which groups are most heavily impacted by the costs, and so on).

6.3.3 MANAGEMENT CONTEXT

Management context significantly influences how the disparate tools of key disciplines can best be integrated to address the multiple dimensions of environmental decisions and, especially, the core question of valuation. Especially salient variables in this management context include the purpose, scope, and authority that frame these decisions. Examinations of present implementation within these basic purpose, scope, and authority variables have revealed significant inadequacies in ecological information that may limit current abilities to optimize integrated valuation decision making, largely due to historic institutional problems. However, systematic application of a template encompassing the basic decision variables may provide a means for identifying potential improvements in integration of valuation at every step in the environmental decision formulation and analysis process. It will be important to focus on institutional considerations in addressing the need for sufficient pertinent ecological data. Such an effort will likely require a broadly based, collaborative effort of decision makers at all levels and specialists from disciplines beyond the core valuation subject areas.

6.3.3.1 Decision Purpose

Environmental management involves a range of decisions with different purposes. These purposes may be differentiated based on the broad management functions they serve, which in part reflects a continuum of quasi-sequential management decision points. This range of purposes and related decisions includes the following:

- *Strategic direction* at the top management level of an enterprise or agency concerning broad goals or directions in land and resource management, regulatory, and restoration programs
- *Program design* decisions to implement major program elements of the strategy

Organizing and Integrating the Valuation Process 139

- *Resource allocation* determinations about emphasis among programs or program subelements
- *Planning decisions* to implement programs at specific administrative units or various geographic scales
- *Project-level* determinations related to implementing programs or plans at a more local site or facility scale
- *Incident response* decisions related to unforeseen discrete environmental damage occurrences, often at a site scale
- *Information system design* decisions regarding research, monitoring, adaptive management and inventories, multiscale assessments, and other ongoing decision support information systems, including those related to process transparency and performance accountability

In addition, both public and private sector decision makers share these sets of varying decision purposes, as each of these types of decisions is dealt with in both the public and private arenas.

6.3.3.2 Decision Scope

Environmental management also involves a range of decisions with differing scopes. Especially salient attributes of decision scope include scale, focus, and goals.

6.3.3.2.1 Scale

Environmental management decisions encompass the effects of human activities on the conditions and operations of natural systems, including the components, structures, and functions of ecosystems. Because ecosystems are manifested in dynamic, nested hierarchical structures and are distributed across landscapes (and seascapes) in patterns determined by characteristic ecological processes acting on them (and functions they perform for one another in a networked fashion), effects of human activities on these components, structures, functions, and processes are likewise manifested at different spatial and temporal scales (Vogt et al. 1996). Examples of different environmental management decisions encompassing varying spatial and temporal scales are shown in Table 6.1.

Although spatial and temporal scales are most commonly discussed as attributes of environmental decision scope with reference to ecological effects, it is important to note that sometimes equally significant analogous spatial-temporal frameworks are employed with respect to economic and cultural effects. Similarly, the distribution of sectoral economic or community social effects of regulatory decisions are integral to government decision making in such programs as the Clean Air Act and the relicensing of dams by the Federal Energy Regulatory Commission (USDOE 1999), as well as by private sector participants in the environmental management decision process.

Some decisions, although made at a single scale, are designed to be integrated vertically with related decisions at other scales. Examples are the Sierra Nevada Framework Assessment, the Great Lakes Ecological Assessment, the Pacific Northwest Forest Plan, the South Florida Ecosystem Restoration Plan, and the Appalachian Regional Assessment. Thus, a key feature of the environmental management context is not only decisions at different scales but also the linkage of decisions at multiple scales.

TABLE 6.1
Spatial and temporal scale of environmental decisions in a valuation context

Spatial and temporal	Near term (< 5 years)	Mid term (5–10 years)	Long term (> 10 years)
International	• UN National Action Plans	• EU CO_2 regimes • UN Water Initiative	• Kyoto Treaty • Iraq-Saudi oil spill • IUCN World Heritage Site designation
National	• Creation of conservation and mitigation banks	• Revision to US Clean Water Act "watershed" rule • US Clean Air Act residual risk regulations	• Revision to Clean Water Act • Benefit–cost analysis of Clean Air Act • Philippines value-based approach to formation of national park
Regional	• US Clean Air Act 305 (b) monitor programs	• Watershed management strategies • CALFED basin plans • USEPA effluent trading policy • Ozone strategic implementation plan • Wildfire management in national parks	• CALFED Water Management Program • Chesapeake Bay Sediment Management Strategy
Local	• Water pollution controls • Air pollution controls • Remedial technology selection	• National discharge permits • Marketable permits for air emissions • Natural resource damage assessment	• Town land use management plans

6.3.3.2.2 Focus

Just as environmental management decisions may be made with reference to a particular geographic or temporal scale, they are often focused on individual environmental factors such as specific species or media like water or air. However, a companion attribute to multiple-scale consideration in defining the scope of environmental decisions is the ecological paradigm of multifactor consideration, or holism. This reflects the management concern for the entire interdependent suite of components in an ecosystem (i.e., their holistic character required in many decisions

TABLE 6.2
Examples of single-factor and multifactor decision focuses

Single-factor focus	Multifactor focus
Endangered species	Toxic substance releases
Clean Air Act particulate standards	Watershed management projects
Water quality standards for total maximum daily loads	Wildland fire management planning
Invasive species controls	Federal Energy Regulatory Commission hydropower relicensing
Migratory Bird Treaty Act assessments	Land and resource management plans for national forests and Bureau of Land Management land units

employing National Environmental Policy Act of 1969 (NEPA) analysis or EPA's Ecological Risk Assessment Framework USEPA (1992) and see Table 6.2).

6.3.3.2.3. Goals

Even though the scales and focuses of particular environmental management decisions may be quite broad (i.e., covering concerns about multiple environmental factors across large geographical areas and considerable time ranges), the decision itself may be directed at only one or a few goals. These goals can vary and include assessments and decisions on ecological conditions, potential of resources for human use, maintenance or restoration of ecosystem functions, and continued or restored provision of ecosystem services such as water quantity and quality.

An example of this is the assessment and decision regarding a toxic release in Lavaca Bay along the Texas coast. The decision involved an assessment directed at identifying the potential adverse effects on ecosystem components; however, because of concentrations of mercury in fish tissue at levels that precluded consumption of the fish by humans, the goal of the environmental decision making considered the risk to human health from fish consumption and the loss of recreational fishing services due to a ban on fishing, as part of the natural resource damage assessment action at the site.

Although many environmental decisions are directed at single goals, such as the example above, many others are directed at multiple goals. This may be explicit, as in multiple-use land management planning in the US Forest Service and the Bureau of Land Management in which environmental analyses are to be framed around providing, in perpetuity, the goal of sustainable uses of specific named resources including rangeland for forage, recreation, timber, wildlife and fish, water, wilderness, and timber.

6.3.3.3 Decision Authority

A third major variable in the management context of environmental management is the legal and policy authorities underlying environmental decision making. The attributes of this variable explicitly include both environmental laws and general management statutes. Together they are aimed at not only effective environmental and resource protection and use, but also efficient, accountable management to achieve them.

6.3.3.3.1 Environmental Laws

Most decisions in the environmental management process are based on statutes, regulations, and agency policy directives adopted incrementally in a piecemeal fashion over the last 4 decades and in response to changing conditions, understandings, and societal concerns (USGAO 1994). Moreover, further elaboration and refinement of these legal and policy criteria for decision making continue to evolve through an even more extensive, dynamic body of administrative hearing and judicial interpretations.

In many cases, these legislative, administrative, and judicial authorities explicitly specify procedural requirements regarding the use of ecological, economic, and social data and associated analytic methodologies, both guiding and bounding the use of these data and methodologies. For example, in the United States, the Endangered Species Act specifically forbids the use of economic effects data in determining the status of such species. Conversely, agency guidelines for development of multiple-use land and resource management plans, which must be done in accordance with NEPA, explicitly require consideration of specific economic and social factors as well as analysis of the differing effects of a range of alternatives, including those that maximize one or more ecological or economic factors. From a procedural mechanism standpoint, the elaboration in statute, regulation, and policy directive under the National Forest Management Act of the establishment of Inter-Disciplinary Teams (IDTs) involving multiple specialists having ecological, economic, and social expertise illustrates the concern for integration in the environmental management arena. The IDT approach has been institutionalized in virtually all significant Forest Service project-level management decisions.

Additionally, as mentioned elsewhere in this volume, the USEPA's environmental risk assessment process specifically requires consideration of commercial and aesthetic values. Similarly, the Montreal Process criteria and indicators include ecological, economic, and social indicators for use in federal management of temperate forests, and parallel federal interagency efforts are underway to develop similar sets of indicators for rangeland, water, and soils (USDA 2003).

6.3.3.3.2 General Management Statutes

In addition to environmental and resource management laws, another set of US statutes is increasingly important to the management context of environmental decision making. These statutes are often referred to as the "management reform statutes of the 1990s" and generally are administered by the US Office of Management and Budget (USOMB) in the Executive Office of the President. Among these reforms is the Government Performance and Results Act (GPRA) of 1993, which requires all federal agencies to prepare strategic plans identifying quantitative outcome-based performance measures and goals that must be linked to annual budgets. GPRA has had a profound effect at forcing agencies to evaluate their programs, and has essentially forced federal agencies to perform valuation as an input to decision making (USOMB 2003).

6.3.4 MANAGEMENT CONTEXT INFLUENCES ON INTEGRATION: CURRENT LIMITATIONS AND OPPORTUNITIES FOR IMPROVEMENTS

Examination of the management context of environmental decision making provides environmental decision makers and specialists a basis for identifying both conditions currently limiting better integration of ecological, economic, and social analysis and opportunities for future improvements to integration that might yield or achieve better decisions and results. These 2 conclusions may be illustrated by focusing on a central issue around which better integration is considered needed: identifying more appropriate methods for applying resource valuation techniques to integrate multiple types of value.

6.3.4.1 Current Limitations on Integration within the Management Context

Better integration of disparate valuation approaches in environmental decision making is currently limited by implementation within the existing management context. The principal limiting feature is the failure to identify more consistently and completely necessary information on ecological conditions on a multiscale, multifactor basis across the numerous programs affecting natural system components, structures, processes, and functions. This same problem exists with economic and social information. But the problem with ecological information is most critical because environmental programs are framed around ecological conditions. Improvements in integration of valuation approaches depend on a chain of considerations that is only as strong as its weakest link. That weakest link is frequently the failure to present relevant ecological information in a form that can be understood and used by decision makers. Recent guidelines (e.g., USEPA 1998) identify values as the basis for conducting ecological risk assessment, and employ the data quality objectives (DQO) process as a means of ensuring that all necessary data of appropriate quality are collected as the basis for decision making.

The inadequacies of ecological information in assessment and planning activities may in part be the result of ecological systems having less organized and focused human constituencies than do economic and social interests. It may also be because these data are often inherently more complex and subject to particular uncertainties and difficulty of precision in expressing linkages among them than data for human-defined systems are. However, whatever the reason, the evidence from numerous government and private analyses repeats the refrain of these critical inadequacies, which makes it much more difficult and in many instances impossible to accomplish optimal integration with economic and social data. The approach presented in this chapter provides a systematic basis for identifying ecological values that can be expressed as "services," making them easier to integrate with social and economic information to support environmental decision making.

For example, even at relatively macroanalysis levels, recent efforts to identify multiscale environmental indicators were found to be significantly hampered by the fact that data for at least a third of these indicators were unavailable (NAS 2000; H. John Heinz III Center for Science, Economics and the Environment 2002). Moreover, this problem has been continually identified by these and other entities and analysts in numerous specific agency and program settings (USGAO 1997, 1999).

The cause for this is rooted in certain key historical artifacts from the development of the management context discussed here, and their enduring influence on its implementation represents an onerous legacy that has been passed down to the present day.

Central among these historical artifacts is the fragmentation of analysis and decision-making authority arising from the piecemeal assignment to different governmental agencies of responsibilities under incrementally adopted statutes occurring over decades, largely based on political circumstances of the times. Because of this, several agencies may have responsibilities for environmental decisions affecting the same ecosystems, their components, or their processes. However, these agencies, having authority independent of one another, often adopt different standards for identifying, distinctive processes for assessing conditions of, and varying standards for making decisions about goals and value trade-offs for these same ecosystems, components, and processes (NCSF 1993; USGAO 1994). Sometimes these disagreements have complicated public stakeholders' efforts to resolve planning and project conflicts because agencies have had to elevate issues internally and been unable to make commitments, as in the case of the multistakeholder Applegate Watershed Partnership's efforts to implement the Northwest Forest Plan and the relicensing of privately operated hydropower dams on federal lands (USGAO 1997).

The fragmentation of environmental decision-making authority among several federal agencies has had especially pernicious effects on the completeness and comparability of different agencies' information on ecosystems and their components, structures, and processes. Although one might imagine that, with multiple agencies being involved, there would be an abundance of good information, the opposite is more often the case. Part of the reason is because some agencies (notably, the USEPA) use compliance data that are difficult to aggregate in a meaningful way to characterize changing ecosystem conditions. However, there are additional complications in the case of water quality data collected under the statutes at the federal level. Many are related to state-based standards that may differ, even when involving the same body of water. Thus, in some instances water quality in the Mississippi River is acceptable in the same spot according to one state, but not according to another.

Similarly, after more than 15 years of an increase in the annual average number of major fires, the US Department of the Interior and the US Forest Service are unable to identify the number of acres with hazardous conditions with less than a 50% error range. Moreover, the data they have at the national scale cannot be related to site-scale data for project identification and assessment (USGAO 2003). Finally, 1 reason why many agencies lack adequate data on ecosystem conditions is because they often do not collect monitoring information as promised or required on the outcome of prior actions because it is not a high priority. This, in turn, makes it difficult to adequately inform adaptive management decisions about whether a change in current or planned management activities is necessary.

Perhaps the most critical underlying problem with ecological information relates to the fact that the boundaries of ecosystems generally transcend the administrative boundaries (functional and geographical) of the agencies. This is largely the result of the fact that the agencies' responsibilities and jurisdictions were established considerably prior to the widespread acceptance of and efforts to apply the ecosystem construct in management, as were many of the major information systems that they established to carry out their responsibilities. In many senses, the inadequacies, lack of comparability, and inconsistencies in agency ecological information systems are rooted in this fundamental "transboundary" circumstance. While agencies may seek to make adjustments in their information systems, the disjuncture is profound and correcting it will be extremely expensive.

One difficulty in correcting it is gaining consensus on the appropriate kinds of information to be applied to different circumstances. For instance, land management agencies have tended to increase their reliance on information gathered and summarized on a watershed basis, in part because of the need for more information to help them manage important water-related issues. However, data collected in this way are considered in many contexts to be less useful in characterizing and predicting future conditions for a wide range of terrestrial functions. Moreover, ironically, it appears that in at least some scales and systems, the utility of information gathered or classified on the basis of the National Hierarchical Framework of Ecological Units — a terrestrially based ecological classification system — has proven to be more accurate in predicting watershed conditions than has information gathered using the US Geological Survey's Hydrologic Unit Codes, a watershed-oriented framework (MoRAP 2001).

The most immediately noticeable difference in agencies' approaches caused by fragmentation of authority is the dissimilar specific procedures and terminologies employed to assess risks to ecological systems. However, while terminologies may differ, the different agencies frequently utilize similar underlying frameworks for their processes and share a generically similar 2-phase evaluation process of problem formulation and analysis.

6.3.4.2 Identifying Opportunities for Better Integration

As a matter of first impression, the most logical and seemingly only way to correct the adverse legacy of fragmented authority and administrative and natural system boundary differences would be to unify authority in the environmental management context under a single agency, such has been periodically recommended by various national commissions. However, it has been noted that a less politically and administratively burdensome alternative approach to such a massive reorganization may be more rigorous, systematic coordinating and streamlining among existing agencies and authorities.

A key sign of convergence is that several agencies aside from the USEPA (especially land and resource management agencies) are increasingly becoming involved in developing new, more explicit, and broader formal risk assessment strategies as the principal focus for addressing new large-scale environmental challenges such as invasive species and wildland fire management. In doing so, they are building on both the USEPA's experience and their own earlier experiences implementing the US Endangered Species Act. The result of this is an increasing cross-fertilization of understandings and visions among environmental professionals within the scientific community as well as within the managerial community. What is needed now is a more organized cross-fertilization at the intercommunity level, bringing both communities together to meet their interconnected challenges jointly.

At some point it may become clear what, if any, changes in laws can help better resolve these technical and managerial challenges. But what changes might be best will only become clear through identifying the limits of what can be accomplished within the existing legal and managerial context. The objective should be defined not as replacing (or even restructuring) the environmental decision process, but as overhauling it. In making progress toward accomplishing this objective, an initial common task is to articulate and present an integrated approach to the problem.

This initial task should be carefully organized and involve several other parties with valuable contributions to make. Eventually, these should include the following:

- Decision scientists who can help bring hierarchy analysis and other tools to the formulation of a more coherent view of the decision process itself
- Enterprise architecture information technologists who can begin assessing and articulating the nature and extent of decision support infrastructure needs
- Program evaluation specialists who can assist in the design of transparency and accountability models that will be needed to provide corollary internal and external test and justification functions
- Financial management experts who can assess resource needs and potential budgetary configurations for meeting them and ensuring fiscal efficiency

But at the outset, there are needs for more extensive, organized discussion about these tasks and the overall strategy they are intended to serve. This discussion needs to involve representatives of professional organizations in the core disciplines and public decision entities involved, as well as industry and the academic and NGO communities, to better frame the technical and institutional issues. It will be particularly important early on to identify and involve key players from the legislative branch of government who may provide valuable insights about issues of importance from that perspective and the kinds of sponsorship resources it may bring to a broader exercise. It is likely only from such a broad range of discussions that not only a workable strategy but also an adequate sense of how to build sufficiently broad support to carry it out can emerge.

6.4 COMPREHENSIVE AND SYSTEMATIC APPROACH TO IDENTIFYING ASSESSMENT ENDPOINTS

6.4.1 Assessment Endpoints

Problem or goal identification should be the first step in environmental decision making. Once this step has been taken, the various decisions at various stages in the process can then be specified, and information necessary to make the decision(s) can be collected and analyzed. Various frameworks have been developed to guide the overall process. The process of identifying assessment endpoints within an overall framework serves to focus data collection to make decisions based on all data of the appropriate type and quality.

Organizing and Integrating the Valuation Process

Existing frameworks of hazardous waste laws, regulations, and guidance documents that provide for environmental protection offer only general concepts regarding which aspects of the environment are to be protected. In many countries, environmental protection includes both ecological and human environments, but these are often addressed as separate issues. For example, the USEPA's "Risk Assessment Guidance for Superfund," volume 2 (USEPA 1989b), equates environmental relevance with ecological relevance. This is significant in that it unambiguously recognizes relevant ecological values (i.e., properties necessary to sustain a healthy ecosystem) as distinct from human use values in the risk assessment process. To avoid confusion, the concept of ecological values, as defined by the USEPA (1998) and Reagan (2002), will hereinafter be replaced by the term "properties." Human health risks are addressed by separate guidance documents, and other social and economic issues may not be adequately addressed because they fall in gaps between regulatory requirements or processes established for decision making.

Norton (1987) and Harwell et al. (1994) provide a philosophical basis for understanding that all values placed on ecological resources are ultimately human values. These can be logically subdivided into 1) goods, services, and qualities that humans value directly; and 2) ecological properties necessary for maintaining a healthy ecosystem. In this context, the first category refers to direct human uses (e.g., hunting, fishing, or timber extraction) and to nonconsumptive considerations (e.g., aesthetic and spiritual values such as wilderness or ritualistic uses by indigenous people), whereas the latter category are those features that are necessary for maintaining the ecosystem (e.g., biological diversity, productivity, and keystone species). While many of these attributes are ecosystem specific, those at the highest levels of ecological organization are common to all ecosystems.

Assessment endpoints based on sustaining healthy ecosystems and incorporating environmental (ecological) values that people care about are influential in making risk management decisions (USEPA 1992, 1998). Identifying as many assessment endpoints as possible is a critical but often neglected step in the valuation process that supports environmental decision making. However, assessment of ecological endpoints is often hampered by lack of funding, as well as time, data, and model constraints. What we present here is a comprehensive and systematic approach to identifying ecological endpoints as the basis for environmental decision making for environmental assessment, planning, or management projects. The approach has the additional benefits of transparency, therefore supporting communication and stakeholder participation, which are key components of any environmental decision-making process.

6.4.2 Problem Identification and Goal Setting

Whether environmental decisions are made to meet future goals or to resolve current problems depends entirely on the circumstances and regulatory context of the issues (see section 6.3). The DQO process (USEPA 2000a) provides useful guidance for framing appropriate problems for risk assessments and other unanticipated environmental problems (e.g., oil spills and releases of hazardous substances). Initial decision making generally involves appropriate technical experts of the various natural resources that may have been affected (e.g., hydrologists, biologists, toxicologists, and risk assessors), regulators, and other decision makers. Depending on the regulatory context or framing of the problem, economists may or may not be involved early in the process, and sociologists and anthropologists may not be involved at all or until it is too late to collect time-critical data. Involvement of the public and other stakeholders is frequently specified by law or regulation, but it may be discretionary. Such omissions are frequently the result of the compartmentalized regulatory framework prevalent in many countries.

Problem identification should involve all potentially affected ecological endpoints. Formulation of the problems and goals to be addressed should, therefore, involve all the appropriate disciplines, including economists and other social scientists, to determine what (if any) relevant issues need to be addressed and to recognize possible information needs to make decisions at some point in the decision-making process. Other stakeholders should be identified early so that their input can be obtaining in framing the problem so that pertinent data can be collected and their issues can be appropriately addressed.

6.4.3 Stakeholder Identification

Stakeholders are individuals or constituencies who have ownership of or an active interest in natural resources and services within or potentially affected by the environmental action, whether it be an oil spill, new industrial facility, or park. Early involvement of stakeholders in any decision process is important in defining the endpoints for valuation in value-based decision making. To be most effective, involvement, endpoint identification, and valuation should be developed within a framework that is accessible to all stakeholders, and it must be transparent in order to elicit participation. Typical stakeholders include the following:

- *Planning entities:* Local and regional agencies, federal programs, corporations, international financial institutions, and the like that are proposing projects or developing programs.
- *Regulatory agencies:* Responsible for planning or regulating and proposing action (e.g., Bureau of Land Management for pipeline construction on federal lands, and USEPA for hazardous waste releases).
- *Directly affected parties:* Local communities and indigenous people (e.g., Aborigines in Australia, Native Americans in North America, and Pygmies in Africa); fisheries cooperatives that use the potentially affected resources.
- *Nongovernment groups (NGOs):* Organizations that have an interest in protecting or restoring particular ecosystems (e.g., the World Wildlife Fund, the National Audubon Society, Greenpeace, and local special-interest groups).
- *Other government units:* Government agencies that have some level of jurisdiction over potentially affected resources or whose resources may be affected by the proposed environmental decision making. Such stakeholder involvement is prescribed for some situations in which the agencies are identified decision makers (e.g., Natural Resource Trustees for natural resource damage assessments [NRDAs] in the United States) but are not required for other decision-making activities.

Organizing and Integrating the Valuation Process 149

Stakeholder involvement is an ongoing process throughout valuation and decision making. Stakeholders may cease to participate once they feel comfortable that either their interests are not being affected or they are being adequately addressed by the process. Others may join the process as they become aware of the problem and the opportunity for involvement. Such participation should be encouraged and documented (see the Case Studies at the end of this book). Once stakeholders have been identified, appropriate venues and schedules are provided to facilitate participation and input.

6.4.4 IDENTIFICATION OF ASSESSMENT ENDPOINTS

Systematically identifying relevant ecological endpoints in assessing ecological risk eliminates the likelihood of reaching trivial conclusions and incurring costly delays. Ecological risk assessments that focus on the most sensitive organism, irrespective of its occurrence at a site or of its importance in the food web, are clearly inappropriate from scientific and logical perspectives. Endpoint identification processes that fail to consider relevant ecological properties at different levels of organization (e.g., animal community, keystone species) in a comprehensive manner have resulted in identification of additional endpoints later in the process, requiring additional data collection and delays in decision making. Documentation of the rationale for selecting particular endpoints and not others is generally brief or totally lacking, but important to maintaining process transparency.

In ecological risk assessments, assessment endpoints are defined as explicit expressions of the environmental values that are to be protected (Suter 1989; USEPA 1992). These endpoints are identified early in an ecological risk assessment to ensure that the assessment focuses on relevant ecological endpoints. Human health risks and other social and economic issues are addressed separately and at different stages in the overall decision-making process. Similar approaches create similar problems in evaluations and planning activities for a broad spectrum of environmental projects, including development projects in the environmental impact assessment process, in conservation and natural resource management planning, and in programmatic planning. This fragmentation of processes creates information gaps and increased uncertainty, often leading to protracted decision making, cost-ineffective decisions, and stakeholder dissatisfaction with the outcome (see Case Studies).

The endpoint identification process described here provides a means of thoroughly documenting the rationale for determining relevant endpoints for each potentially affected ecosystem, irrespective of other considerations. These ecological endpoints at all levels of ecological organization provide a basis for determining site- or area-specific values that can serve as goals for planning or endpoints to be considered in development or environmental problem solving (e.g., planning for remediation or restoration).

The original idea for a comprehensive and systematic endpoint identification approach grew out of discussions at the 1993 Pellston Workshop on Sustainable Ecosystem Management, sponsored jointly by the Ecological Society of America (ESA) and the Society of Environmental Toxicology and Chemistry (SETAC). While there was no final consensus on the fundamental question "What do we want to sustain?" there was general agreement that the endpoints to be protected and sustained needed to be identified, and that ecosystem endpoints at all levels of ecological organization would need to be addressed.

The general approach to determining relevant ecological properties was developed further as part of the selection of assessment endpoints for the ecological risk assessment at the Lavaca Bay Superfund site and presented in conceptual form at the 1996 SETAC annual meeting (Reagan et al. 1996). It has since been used successfully in a variety of ecological risk assessments in the United States, for a river basin affected by mine waste disposal in Papua New Guinea, for national park planning in the Philippines (see Case Study 1), and for NRDAs. Reagan (2002) describes in detail the value-based process for identifying assessment endpoints for ecological risk assessments.

Because the process can incorporate endpoints identified in other natural resource activities (e.g., ecosystem management, natural resource damage assessment, environmental impact assessment, and ISO 14000 compliance), it provides a consistent basis for integration among related processes. The stepwise approach also permits a comprehensive and systematic means of addressing all possible endpoints and documenting how each was considered. Because the process reduces the use of scientific jargon and complex scientific concepts, it provides a basis for stakeholder understanding and involvement in the processes of identifying ecological endpoints, planning, assessment, and management decision making.

Determination of assessment endpoints in ecological risk assessments begins by identifying the universe of potential ecological values to be protected (USEPA 1998), then proceeds to identify site-specific properties that embody these values. The process occurs in 2 stages:

- Identify ecological properties necessary to sustain a healthy ecosystem.
- Identify the ecological endpoints directly valued from a human perspective associated with the ecological resources of the ecosystem under evaluation.

6.4.4.1 Identifying Relevant Ecological Properties

The identification of ecological endpoints (e.g., nutrient cycling in an estuarine system) begins by identifying essential properties common to all ecosystems and progresses to a consideration of properties pertinent to the regional or area-specific ecosystems of interest. This progression is hierarchical and scientifically based, thus providing an objective means of determining which components of the ecosystem are potentially relevant.

Recognizing the ecosystem as the appropriate context for evaluating attributes at all levels of ecological organization is consistent with current principles of ecosystem management and with our current understanding of ecosystem organization (Kaufmann et al. 1994; Reichman and Pulliam 1996; Boyce and Haney 1997). "Ecological relevance," as used here, refers to the properties necessary to sustain ecosystem components and functions.

The process of identifying ecologically relevant endpoints (Stage 1 above) consists of the following steps:

- Identifying ecological properties common to all ecosystems
- Identifying functional components of regional ecosystems
- Developing the food web of potentially affected ecosystems
- Determining ecologically relevant attributes of the functional components of these ecosystems, and stating the ecologically relevant properties in common language

Once these ecological properties have been determined, assessment endpoints relevant to the identified project or problem are developed. Site-specific endpoints are then selected by identifying those that are relevant to the problem or project (e.g., the susceptibility of the endpoints to site-related environmental stressors in a risk assessment, and features to be sustained or preserved in resource management).

In broad terms, an ecosystem can be defined as the habitats, both aquatic and terrestrial, of the site or area of interest. Sustaining a healthy ecosystem is the ultimate ecological value to be protected; however, a variety of other ecological properties and endpoints must also be considered.

6.4.4.1.1 Step 1: Properties Common to All Ecosystems
The approach for identifying ecological properties proceeds hierarchically from the fundamental objective of preserving a healthy ecosystem to the identification of relevant characteristics common to all ecosystems. While there is no generally accepted definition of what constitutes a healthy ecosystem, relevant ecosystem attributes are as follows:

- *Biological diversity (biodiversity):* Ecological structure in terms of components; can include species, community or habitat, and genetic diversity.
- *Structure and function:* These properties describe the pattern of organization of ecosystem components; not only are all of the ecosystem components present, but they also interact according to organizational principles typical of that ecosystem.
- *Cycling and transport processes:* Appropriate nutrient and energy dynamics must be operating within the ecosystem to maintain balanced populations at levels typical of a healthy ecosystem. For an ecosystem to function normally, the flow rates of energy and nutrients (e.g., primary productivity and decomposition) should fall in a range typical of that type of ecosystem. Disruption of flow rates could lead to accumulation of detritus, reduction of energy inputs, or loss of top predators that could alter energy flow patterns and change ecosystem structure (McNaughton 1978).

These properties express the components, patterns of organization, and process rates, based on fundamental ecological concepts expressed in terms that can be evaluated. Because they are features relevant to all ecosystems (Odum 1993), they are relevant ecological attributes. All other ecologically relevant issues are subsets of these valued characteristics common to all ecosystems.

6.4.4.1.2 Step 2: Functional Components of the Regional Ecosystem
Because food webs provide essential structural organization in ecosystems (Gal-lopin 1972) and because all organisms in an ecosystem are part of the food web,

the food web concept is used to identify basic functional components of potentially affected ecosystems.

Food webs are typically composed of 3 basic trophic categories:

- *Producers:* organisms that manufacture food from inorganic compounds by photosynthesis or chemosynthesis (e.g., green plants and chemosynthetic bacteria)
- *Consumers:* organisms that ingest other organisms (e.g., animals that consume plants or other animals)
- *Decomposers:* organisms that derive their nourishment from dead organic matter (e.g., fungi and bacteria)

The 3 fundamental food web categories need to be categorized into functional groups (components) based on a general knowledge of the species present in regional ecosystems. Consumers are grouped in feeding guilds (e.g., organisms that obtain their food in a functionally similar way). For example, many organisms in a forest eat insects, but those that forage in the canopy perform a different functional role than those that feed in the litter of the forest floor. Therefore, food webs based on feeding guilds facilitate the identification of critical ecosystem functions and the interrelationships among guilds that may affect other ecosystem properties.

By using functional rather than taxonomic groups, we are able to determine broad functional aspects of the ecosystem and important interrelationships among components, while avoiding the impossible task of determining the ecological relevance of each species. To illustrate this approach, consider the food web of one of the most complex ecosystems on earth, a tropical rain forest (Table 6.3). Functional components were developed based on food, foraging location, and food habits of species present in the forest. The essential point is that all species fit into one or more of the functional components, and thus all species are addressed at least at this functional level.

Stakeholder involvement is crucial to the identification of environmental endpoints. Ultimately, the effectiveness of an assessment or planning project depends on how it improves the quality of management decisions, and these decisions should be based on endpoints to be protected. From the standpoint of endpoint identification, it is significant that even a complex ecosystem such as a tropical rain forest, which is composed of thousands of plant and potentially millions of animal species, can be comprehensively represented in functional terms by less than 20 functional components. It is also noteworthy that despite the greatly different taxonomic components, all forest ecosystems have essentially the same functional groups.

6.4.4.1.3 Step 3: Functional Food Web
The functional components described in Step 2 define the general range of feeding preferences and locations (strata) in the New Guinea rain forest (Figure 6.1). The arrows in the food web diagram indicate the direction of flow of energy and nutrients through the food web. Dashed lines indicate the recycling and flux of energy and nutrients as the result of decomposition processes. The diagram shows major pathways; others may be present, but are either insufficiently understood or relatively insignificant.

Organizing and Integrating the Valuation Process

TABLE 6.3
Functional components of the tropical rain forest ecosystem in New Guinea

Basic trophic category	Functional component
Producers	Ground layer
	Understory trees
	Canopy trees
	Lianas and epiphytes
	Mycorrhizal fungi (enhancers)
Consumers	**Herbivores**
	Terrestrial frugivores and granivores
	Arboreal frugivores
	Folivores and browsers
	Nectarivores
	Carnivores and omnivores
	Intermediate and small predators
	Omnivores
	Top terrestrial predators
	Top arboreal predators
Decomposers	Detritivores and scavengers
	Chemical decomposers

6.4.4.1.4 Step 4: Ecologically Relevant Attributes
While feeding relationships are relevant characteristics of each functional component of the terrestrial food web, each component may have additional ecologically relevant attributes that define its overall value. For many functional components, the

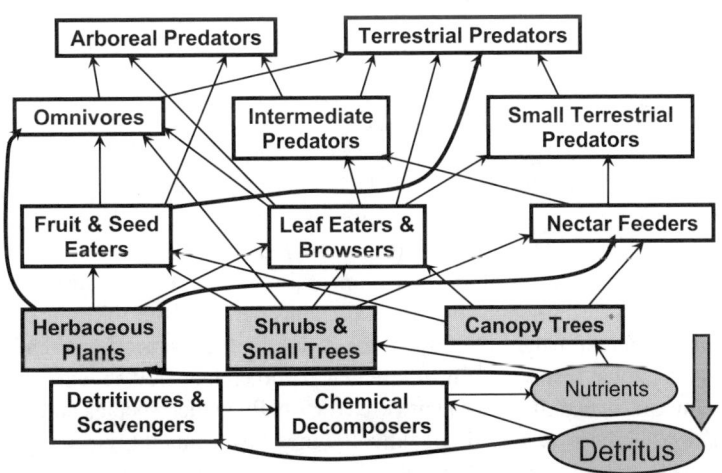

FIGURE 6.1 Food web for a forest ecosystem (modified from Reagan 2006).

nontrophic attributes are at least as important as their role in nutrient and energy transfer through the food web.

Relevant attributes of the ecological components of the New Guinea rain forest ecosystem are defined below:

Food: Source of energy and nutrients for ecosystem components.

Habitat: Shelter or structural support for other organisms.

Energy and nutrient fixation: The process of converting inorganic chemicals to organic compounds that can provide energy and nutrients to other living components of the ecosystem.

Decomposition: The breakdown of nonliving organic matter, recycling nutrients and preventing an accumulation of nonliving organic matter that would interrupt energy- and nutrient-cycling processes.

Propagule dispersal: The distribution of seeds and spores from their origin to other locations. The process is important for recolonization and natural revegetation following disturbance.

Pollination: The cross-fertilization by nectar- and pollen-feeding animals is the sole means of reproduction in many plant species.

Predation: The killing and consumption of other animals by carnivores, including top predators, insectivores, and parasites.

Control: Either "top-down" or "bottom-up" effect on the structure (composition and abundance) or function of the ecosystem.

In Step 4, the various attributes are determined for each of the functional components of the ecosystem and arranged in a table (Table 6.4), where each functional component has at least one attribute. While some of these attributes could be considered more important than others, the table captures all ecosystem components at a functional level and considers the characteristics of each, thus providing a means of describing ecological attributes and subsequently identifying ecological endpoints.

6.4.4.2 Identifying the Human Values of Ecosystem Components

As previously indicated, the second stage in a comprehensive and systematic identification of assessment endpoints in ecological risk assessments considers human values associated with ecological resources. Ultimately, the effectiveness of an assessment depends on how it improves the quality of decisions. Managers are more willing to use risk assessment as the basis for making decisions if the risk assessment considers ecological endpoints relevant to stakeholders (USEPA 1998). Stakeholder relevance is also critical for other environmental planning and management activities. Therefore, environmental projects must consider both the relevant ecological attributes and societal values of ecosystems. These include formally recognized and protected ecological features such as threatened and endangered species, and recreationally important (e.g., game) species plus regionally valued species and resources. Identification of these values should involve input from managers, assessors, and the public. It is particularly important to consult local citizens and understand the values they place on ecological resources. Where the public includes members of indigenous (e.g., First Nations, or Native American Indian tribal) communities, the

TABLE 6.4
Ecologically relevant attributes of the functional components identified for a tropical rain forest ecosystem in New Guinea

Functional components	Ecological attributes						
	Food	Habitat	Primary Production	Pollination	Seed Dispersal	Decomposition	Control
Ground layer	X	X	X				
Understory trees	X	X	X				
Canopy trees	X	X	X				
Lianas and epiphytes	X	X	X				
Mycorrhizal fungi			X				X
Terrestrial frugivores and granivores	X				X		
Arboreal frugivores	X				X		
Folivores and browsers	X						
Nectarivores	X			X			
Intermediate and small predators	X						
Omnivores	X						
Top terrestrial and arboreal predators							X
Detritivores and scavengers						X	
Chemical decomposers						X	

endpoints identified may include a broad range of ecological components and human use values such as food, medicine, and ritualistic purposes.

Determining societal values involves a systematic search of all pertinent information sources, including but not limited to

- reviewing existing information regarding rare and threatened species and conducting interviews with regional experts,
- reviewing information on commercially and recreationally important resources, and
- obtaining input from the public (including indigenous communities) on natural resource values.

From an international perspective, societal values can include formally recognized and protected ecological entities such as threatened species. However, local human uses may be in conflict with the values placed on a resource by the international demand for the same species. For example, local communities may wish to continue to extract resources unsustainably from natural ecosystems, which is in conflict with the concern over protection of global biodiversity.

The comprehensive identification of general endpoints provides a comprehensive summary of the values of the ecosystem. The process of selecting site-specific endpoints requires an understanding of which functional components or species are potentially endpoints to be sustained or potentially affected by planned or unanticipated disturbance.

6.4.5 Data Collection

In the value-based, decision-focused framework, identified values represent the endpoints to be protected or sustained, and decisions are made regarding how this will happen. In ecological risk assessments, values are the assessment endpoints. In environmental impact assessments, they are the affected resources. In natural resource damage assessments, they are the potentially affected resources, and the services they provide, which are the endpoints to be restored. In other environmental management activities, such as conservation or park management, values represent the set of goals to be sustained. Using a comprehensive system for identification of relevant ecological attributes provides a means of maximizing data usability among these different environmental activities.

Informed decisions are made based on relevant information concerning the endpoints to be protected or sustained. Information can come from a variety of sources, including published literature, unpublished reports, and location or site-specific data. Approaches such as the Data Quality Objectives (DQO) process (USEPA 2000a) present a means for ensuring that all necessary data of the appropriate type and quality are collected to inform decisions. The process consists of the following 7 steps:

- State the problem.
- Identify the decision to be made.
- Identify inputs to the decision.
- Define the study boundaries.
- Develop a decision rule.
- Specify limits on decision errors.
- Optimize the design for obtaining data.

Step 1 is to develop a clear concise statement of the problem or goal for which decisions are required. Subsequent steps identify the types of decisions to be made, information needs, and available information, focusing on the types and quality of data necessary to adequately support decision making. Data gaps are then filled by collecting all (but only) data necessary to support decision making. If properly implemented, analytical and statistical methods and limits on decision errors are developed in advance of sampling, in order to design an adequate data collection program. Both sampling errors and measurement errors are combined into a total study error that is directly related to decision error.

Because not all data collection outcomes can be anticipated, one possible decision to be made is that more data on a specific topic are needed to support decisions

Organizing and Integrating the Valuation Process

on a specific resource (e.g., plant productivity). Such tiered approaches to data collection are likely necessary when dealing with complex sites or problems.

The DQO process thus ensures that all data collected are usable for decision making, and that the process is documented and transparent to regulators and the public. Use of a comprehensive framework for values identification maximizes data usability among environmental activities. The process is appropriate for all types of data collection, including measurements of physical resources (e.g., hydrology, geology, and air quality), biological resources (e.g., endangered species and biodiversity), and socioeconomic values (e.g., recreational services and human health risk).

6.4.6 USING ECOLOGY AND ECONOMICS TO IDENTIFY ENDPOINTS FOR VALUATION AND PROVIDE INFORMATION FOR DECISION MAKING

Any meaningful framework for integrating economics and ecology to aid decision making must begin at the problem or goal identification stage. Before decision makers can decide whether to go ahead with a project, they must understand what is valued in the ecosystem and how the values associated with the resource will change as a result of the decision. However, the specific information needed for a valuation of an ecosystem service is usually very context specific, that is, the value that needs to be measured or considered will differ depending on the specific question being asked. Thus, the processes of problem formation and value identification are intricately related and will likely need to be determined using an iterative process. In the following section, we discuss key places where integration among ecologists and economists can improve the final information that is provided to decision makers concerning the values associated with alternative possible decisions.

Information needs should adequately inform the environmental decision-making process. Thus, collection of data to fill gaps should support decision making in a framework such as the DQO process. Following this approach, documents describing data collection and use (e.g., work plans or sampling and analysis plans) should clearly indicate how data to be collected support the decision and, to the extent possible, how the results of data collection will be used to make decisions (e.g., specify decision criteria, statistical methods, and levels of significance).

Once assessments have been performed and decisions have been made regarding ecological endpoints, economists contribute to the decision-making process by providing costs of various remedial options. Monetary damages can also be calculated for natural resource injuries as a basis for supporting restoration in a natural resource damage assessment context. Economic valuations are also necessary for environmental planning activities such as park establishment (see Case Study 1) or environmental impact assessments associated with development projects. For example, in the CALFED Bay–Delta Program (Case Study 6) environmental managers must balance a diverse set of demands from a diverse set of stakeholders on a limited quantity of water. Ideally, valuation into a common metric would allow the comparison of the value of water for agriculture, urban drinking water, endangered species protection (e.g., Chinook salmon), and other uses. Economic valuation adds an essential reality to environmental planning and management activities. There are many places where the integration of economics with ecology will require careful attention to the type of information available for use.

1. *Spatial and temporal scales*: To undertake valuation studies, ecological information needs to be at spatial and temporal scales that can be understood by people and that are ecologically relevant. Likewise, it needs to be aggregated enough to provide meaningful differences to people.
   - It is difficult for people to place a value on a very small change in risk, or even to envision a very small risk itself (e.g., what one is willing to pay to decrease his or her risk of air quality induced mortality from 1 in 500 000 to 1 in 2 000 000 is difficult to assess.) Thus environmental changes on a very small scale are difficult for the public to value. Their valuation is complicated by the fact that measures such as risk numbers, quantification of chemical measures, nutrient levels, air quality indices, and so on are usually not very meaningful to people when accessing the value of a change in environmental quality.
   - At the same time, it is critical for economic valuation methods that the alternatives be accurate and believable from an ecological (scientific) perspective. If not, valuations may be biased. For example, is a "pristine" body of water even achievable given the state of the water body and surrounding environment?
   - The links between properties of ecosystems that people notice or value are not well understood by economists or those who are being studied (e.g., water quality and perceptions). For example, people do not notice increases in heavy metals in local swimming sites, and need to be informed of the increased risk in order to value any risk reduction from decreasing heavy metal loads. Ecologists and economists need to work together to better link people's perceptions of their environment with the attributes they value.
2. Range of alternatives considered:
   - Economists often want information on a continuum of stressor levels or distinct alternatives (e.g., a regulation may not change a water body from "boatable" to "swimmable"). Thus, when using a typical "water quality ladder" based on these definitions, there would be no benefit from an increase in water quality unless a threshold is crossed that would impact an environmental attribute for which people have expressed a value.
   - For valuation of a local water quality project, researchers need information on both the baseline situations (current condition, the condition in the future if no project is undertaken, etc.) and the improved (with project) condition.
   - For a decision regarding the amount of land or habitat set aside for a wildlife viewing area, researchers require a relationship between size of habitat and population levels.
   - For ex post-facto damage assessment (such as done in NRDA cases), economists need to know the key information on the preinjury and service levels

as well as the postinjury state of the resource. Thus, assessments of current fish stocks, wildlife populations, habitat, endangered species counts, recreational fishing intensity, and so on are needed. Other data needs include but are not limited to the following:
- Estimates of the duration and severity of the injury over time
- The degree of uncertainty associated with alternative outcomes
- Impacts to other ecosystems, for example injury to an important flyway affects bird populations elsewhere (extent of the market concept)

3. Integrating and accounting for uncertainty across valuation and ecological *risk assessment*:
   - More work on integrating uncertainties across ecological risk assessment and economic valuation is critical. Valuations are estimates from models and therefore subject to standard statistical uncertainty (e.g., sampling error). The valuations also may be sensitive to model specification, and a variety of other sensitivity analyses can be an important tool to integrating the uncertainty present in both the ecological risk assessment and economic valuation methods.
   - Individuals poorly understand risks and trade-off amongst risks. How to better capture consequences of risk and portray them in the context of valuation studies and ultimately to decision makers? Ecological risk assessors and economists should combine and integrate their expertise to develop information pertinent to the values used by environmental managers to make decisions; there is much to be gained from working together.

4. Integrating and accounting for ecological irreversibility in ecosystems and valuation: Irreversibility and adaptive management approaches are especially important when ecosystems are being considered. While Chapter 3 contains a more thorough discussion of the tools appropriate for economic and ecological assessment when irreversibility and the potential for learning with investment are considered, here we note the data needs that the use of these tools would involve.
   - Consider a conservation organization deciding between buying land parcel A or land parcel B (with a fixed budget); parcel A has more species on it, but B has an endangered species whose loss is looming. What is the value of this irreversible loss of a species?
   - In some instances, buying a parcel of land will allow one to find out the value of that type of habitat for preserving a (poorly understood) keystone species. What is the value of reducing uncertainty?
   - Distinctions should be made by preserving areas with unique combinations of valued species and areas necessary for the survival of an endangered species. While the value of the genetic information coded in the genome of a species is not easily valued in monetary terms, its loss may affect other species in an ecosystem, overall ecosystem resiliency, and potential human use services (e.g., medicinal values). The loss of a species through extinction is irreversible.

6.5 INFORMING THE VALUATION PROCESS: SUMMARY

As discussed in this chapter, the context in which a decision maker operates will determine the various types of information that will be utilized when he or she is making a decision. Decision making in the face of uncertainty is the status quo for decision making regarding ecosystems and the environment. In both private and public decision making, many ecosystem services do not carry prices that are reflected in the market, and are therefore usually undervalued in a monetary-based decision framework. For example, a benefit–cost analysis with incomplete accounting of ecosystem services will lead to undervaluing the total benefits of an action. However, in many cases, even nonmonetary valuation measures are very uncertain. Decision makers are often faced with great uncertainty as to the consequences of a decision action, as well as the benefit of the decision when the outcome is well understood. This chapter did not attempt to propose a new framework to systematize the uncertainty brought into the analysis from using various valuation techniques. It merely discusses the role of valuation in decision making, and places it in the proper conceptual context for different decision types. This chapter also points out the influence of the temporal and spatial scales of a decision on the decision process and context.

The obstacles that stand in the way of integration are historical, institutional, and often simply artifacts of the decision context. The decision framework is often incapable of handling an interdisciplinary collaboration at the problem formulation, analysis, or risk characterization stage of the risk assessment process. Nevertheless, it has been shown that early and regular coordination with the social sciences during the risk assessment process can lead to valuation information that is better able to inform the decision maker. Simply promoting a common understanding of terminologies between disciplines allows economists and ecologists to converse in a meaningful way at the very beginning of the assessment process. Having the ability to match the ecological attributes being assessed with the endpoints that are of importance to the valuation and decision process is a great advantage. Both public and private entities need to have the flexibility to adjust and adapt throughout the assessment process.

The main contribution of this chapter has been to point out the gains that may be had from the integration of the social sciences (particularly economics) in the ecological risk assessment process. Better coordination and planning between the disciplines can lead to better, more useful valuation information for the decision maker. The existing frameworks that provide guidance to the ecological assessment process need not be discarded, but simply amended to include parallel communication with the social sciences.

As demonstrated in the previous chapters, there are many differences and similarities between the various types of valuation methodologies. These methods are context specific, and each has the potential to contribute additional information to the environmental decision maker in its own way. However, this additional information must be used with caution. As with many other inputs to the decision process, valuation information can often contain a large amount of uncertainty. Just as inputs to the valuation process can be uncertain, the valuation process itself can be very sensitive to choices in assumptions, data, and methods. Making the valuation process as transparent as possible allows reproduction of the results and greater insights into the decision-making process in which it is being used. As environmental decision-making processes become more quantitative, the demand for valuation information increases. The valuation approaches presented throughout this book all have the common goal of attempting to assign quantitative information to an ecological resource. It is important that decision makers understand the basis as well as the strengths and weaknesses of the information they are given.

The methods used in valuing ecological resources and environmental changes will continue to evolve given the correct conditions. The trend toward collaboration between disciplines will help to increase the rate at which existing methods are improved and new methods are developed. Additionally, increased availability of data throughout all phases of the valuation process is an important ingredient for this continued evolution. In all likelihood, environmental decision making will continue to become more quantitative in nature, and the role of valuation in this process will continue to expand.

REFERENCES

Boyce MS, Haney A. 1997. Ecosystem management: applications for sustainable forest and wildlife resources. New Haven (CT): Yale University Press.

Gallopin GC. 1972. Structural properties of food webs. In: Patten BC, editor, Systems analysis and simulation in ecology. New York (NY): Academic Press, p 241–282.

Harwell M, Gentile J, Norton B, Cooper W. 1994. Issue paper on ecological significance. In: Ecological risk assessment issue papers. Risk Assessment Forum. Washington (DC): USEPA, p 2.1–2.49.

H. John Heinz III Center for Science, Economics and the Environment. 2002. The state of the nation's ecosystems. Washington (DC): Cambridge University Press. 288 p.

[ISO] International Organization for Standardization. 2004. ISO 14001 environmental management guide. http://www.iso.org/iso/en/CatalogueDetailPage.CatalogueDetail?CSNUMBER=31807&ICS1=13&ICS2=20&ICS3=10. 23 p.

Kaufmann MR, Graham RT, Boyce DA Jr, Moir WH, Perry L, Reynolds RT, Bassett RL, Mehlhop P, Edminster CB, Block WH, Corn PS. 1994. An ecological basis for ecosystem management. Technical Report RM-246. Washington (DC): USDA Forest Service.

Khanna M. 2001. Non-mandatory approaches to environmental protection. J Econ Surveys 15:291–324.

McNaughton SJ. 1978. Stability and diversity of natural communities. Nature 274:251–253.

[MoRAP] Missouri Resource Assessment. 2001. Report of the Missouri Resource Assessment. Columbia (MO): MoRAP.

[NAPA] National Academy of Public Administration. 1995. Setting priorities, getting results: a new direction for EPA. Washington (DC): NAPA.

[NAS] National Academy of Sciences. 2000. Ecological indicators for the nation. Washington (DC): National Academy Press. 164 p.

[NCSF] National Commission on Sustainable Forestry. 1993. Agenda for sustainable forestry research. Washington (DC): NCSF.

[NRC] National Research Council. 1983. Risk assessment in the federal government: managing the process. Washington (DC): National Academy Press.

Norton BG. 1987. Why preserve natural variety? Princeton (NJ): Princeton University Press.

Odum EP. 1993. Ecology and our endangered life support systems. 2nd ed. Sunderland (MA): Sinauer.

[PCCRARM] The Presidential/Congressional Commission on Risk Assessment and Risk Management. 1997a. Framework for environmental health risk management. Vol. 1. Final report. Washington (DC): PCCRARM.

[PCCRARM] The Presidential/Congressional Commission on Risk Assessment and Risk Management. 1997b. Risk assessment and risk management in regulatory decision making. Vol. 2. Final report. Washington (DC): PCCRARM.

Reagan D, Campbell T, Gribben K, Beacham J, Cardwell R, Volosin J, Kathman R. 1996. A comprehensive approach to selection endpoints for ecological risk assessments. 11th Annual Meeting of the Society of Environmental Toxicology and Chemistry, Washington, DC, November 15–21, 1996.

Reagan DP. 2002. Determining values: a critical step in assessing ecological risk. In: Paustenbach D, editor, Human and ecological risk assessment: theory and practice. New York: John Wiley, p 1069–1098.

Reagan DP. 2006. An ecological basis for integrated environmental management. Hum Ecol Risk Assess 12:819–833.

Reichman OJ, Pulliam HR. 1996. The scientific basis for ecosystem management. Ecol Appl 6:694–696.

Stahl RG, Bachman R, Barton AL, Clark JR, deFur PL, Ells SJ, Pittinger CA, Slimak MW, Wentsel RS, editors. 2001. Risk management: ecological risk-based decision making. Pensacola (FL): SETAC Press. 192 p.

Suter GW II. 1989. Ecological endpoints. In: Warren-Hicks W, Parkhurst BR, Baker SS Jr, editors, Ecological assessment of hazardous waste sites: a field and laboratory reference document. EPA60013-89/013. Corvallis (OR): Environmental Research Laboratory, p. 2–28.

[USDA] US Department of Agriculture. 2003. Montreal Process criteria and indicators. http://www.fs.fed.us/institute/lucid/montreal_process_criteria_and_indicators.htm.

[USDOE] US Department of Energy. 1999. An assessment of costs associated with hydropower relicensing decisions. Idaho Falls (ID): Idaho National Laboratory.

[USEPA] US Environmental Protection Agency. 1989a. Supplemental risk assessment guidance for the Superfund program. Part 1: guidance for public health risk assessments. Part 2: guidance for ecological risk assessments. Region I. EPA/901/5-89/001. Boston (MA): USEPA.

[USEPA] US Environmental Protection Agency. 1989b. Risk assessment guidance for Superfund: environmental evaluation. Vol. 2. Interim final. EPA 540/1-89/001A. Washington (DC): USEPA.

[USEPA] US Environmental Protection Agency. 1990. Document of ecological risk assessment consultation. EPA-SAB-EPEC-90-LTR-005. Washington (DC): USEPA, Science Advisory Board.

[USEPA] US Environmental Protection Agency. 1992. Framework for ecological risk assessment. EPA/630/R-92/001. Washington (DC): USEPA, Risk Assessment Forum.

[USEPA] US Environmental Protection Agency. 1997. Ecological risk assessment guidance for Superfund: process for designing and conducting ecological risk assessments. EPA 540-R-97-006. Washington (DC): USEPA, Office of Solid Waste and Emergency Response.

[USEPA] US Environmental Protection Agency. 1998. Guidelines for ecological risk assessment. EPA/630/R-95/002F. Washington (DC): USEPA. 188 p.

[USEPA] US Environmental Protection Agency, 2000a. Guidance for the data quality objectives process (QA/G-4). EPA/600/R-96/055. Washington (DC): USEPA, Office of Environmental Information. 100 p.

[USEPA] US Environmental Protection Agency. 2000b. Guidelines for preparing economic analyses. EPA 240-R-00-003. Washington (DC): USEPA, Office of the Administrator. 206 p.

[USEPA] US Environmental Protection Agency. 2000c. Toward integrated environmental decision making. EPA-SAB-EC-00-011. Washington (DC): USEPA, Science Advisory Board.

[USEPA] US Environmental Protection Agency. 2002. A framework for the economic assessment of ecological benefits. Washington (DC): USEPA, Ecological Benefit Assessment Workgroup, Social Science Discussion Group, Science Policy Council. 185 p.

[USEPA] US Environmental Protection Agency. 2003. Integrating ecological risk assessment and economic analysis in watersheds: a conceptual approach and three case studies. National Center for Environmental Assessment. EPA/600/R-03/140R. Cincinnati (OH): 392 p.

[USGAO] US Government Accountability Office. 1994. Ecosystem management: additional testing needed for a promising approach. GAO-RCED-94-111. Washington (DC): USGAO.

[USGAO] US Government Accountability Office. 1997. Forest Service decision-making: a framework for improved accountability. GAO-RCED-97-71. Washington (DC): USGAO.

[USGAO] US Government Accountability Office. 1999. Ecosystem management, western national forests: catastrophic wildfire threatens resources and communities. GAO-RCED-99-65. Washington (DC): USGAO.

[USGAO] US Government Accountability Office. 2003. Wildland fire management additional actions required to better identify and prioritize lands needing fuels reduction. GAO-03-805. Washington (DC): USGAO. 60 p.

[USOMB] US Office of Management and Budget. 2003. The performance reference model. Vol. 1, Vers 1 release document. Washington (DC): USOMB.

Videras J, Alberini, A. 2000. The appeal of voluntary environmental programs: which firms participate and why? Contemp Econ Pol 18:449–461.

Vogt KA, Gordon JC, Wargo JP, Vogt DJ, Asbjorsen H, Palmiotto PA, Clark HJ, O'Hara JL, Keaton WS, Patel-Weynand T, Larson B, Tortoriello D, Perez J, March A, Corbett M. 1996. Ecosystems: balancing science with management. New York (NY): Springer.

7 Synthesis, Recommendations, and Conclusions

Randall J.F. Bruins, Wayne R. Munns, Jr., Larry Kapustka, and Ralph G. Stahl, Jr.

CONTENTS

7.1 Conclusions and Challenges 165
7.2 Recommendations and Research Needs 171
7.3 Moving Forward 172
References 172

7.1 CONCLUSIONS AND CHALLENGES

Conclusion Statement 1. Integrating socioeconomic values with the estimation of potential ecological risks is likely to improve environmental decision making. Socioeconomic values will be dependent on a host of factors, not the least of which will be the diversity in human wants and needs, but this should not deter their inclusion in the decisional process. There is a role for valuation in the decisional process whether decisions are being made in the public or private sectors.

This was a central belief that was shared by many on the workshop steering committee and, fortunately, by participants of the workshop. As a conclusion, it may be common sense, but there's little doubt that effective management and protection of ecological resources depend on well-informed environmental management decisions, which take into account both human values and ecological complexity. Human and other values are implicit in ecological risk assessments, because ecological assessment endpoints reflect environmental management objectives (USEPA 1995, 1997; see also Chapter 2). While in some regulatory contexts these objectives are legally established, very often they are determined based on the values of the risk assessors, decision makers, stakeholders, or society at large. They may or may not be articulated or written, but they are present and help to shape the final decision whether or not it is risk based. While the inclusion of a broad set of values increasingly is seen as a goal for public and private decision making, risk assessors may find this difficult to accomplish in practice. One reason this remains difficult is the absence of a relatively simple method for specific inclusion of valuation (and the underlying analysis) into the ecological risk assessment process. Perhaps just as likely is that, as we have noted in the various chapters of this book, most risk assessors, ecologists, and other scientists seem to have an only rudimentary understanding of valuation, economic or otherwise.

In this workshop we did not seek to develop a "how to" manual on economic or ecological valuation methods that could be used by ecological risk assessors, legal experts, or others. Among us we did, however, engage in open dialog on how valuation should and could be incorporated more formally into ecological risk assessments, and thereby into environmental decision making in the public and private sectors. As noted earlier, the workshop participants, as others before us have done, reached general agreement that the incorporation of valuation would improve ecological risk assessment (USEPA 2005).

Throughout the workshop, and continuing afterward as we developed and edited this book, was the concern that ecologists and economists speak (then and now) a very different language. We had hoped to develop a "Rosetta stone" of sorts during our interactions at the workshop, but we could not, and that is something that will have to be done by others in future venues. All is not lost, however, as noted in 2 of the case studies included in this book, one from CALFED and one from Subic Bay (Case Studies 6 and 1, respectively). At least in these 2 instances, highly diverse groups were faced with the need to undertake the estimation of environmental benefits (broadly defined in both cases) or to include valuation in risk-based decision making. In each case, these groups were able to accomplish this task even in the absence of a formal framework, guidance document(s), or a manual. Reviewing these cases during the workshop was very beneficial to the participants, and thus we have included them in this book. We believe these 2 cases in particular, and the others we have included, will be beneficial for economists, risk assessors, and decision makers to review and thus glean those elements that lead to successful resolution of these difficult problems. We caution, however, that "one size does not fit all," and some of the approaches taken at Subic Bay or by CALFED may not be applicable for some situations. Only after our experience base has grown through the application of the various approaches outlined in this book and elsewhere, will we be able to improve the use of valuation in environmental decision making.

Some of the basis for the conclusions and recommendations we provide in this chapter arose during discussions of questions such as "What made the individuals take valuation into account?" and "Was it due to a regulatory mandate, public outcry, or some other reason?" This often led us to the simple conclusion that many environmental decisions are context driven, seemingly inconsistent across geographies and regulatory programs, and not driven solely by science and technical information. In contrast to the 2 case studies we cite above, the ability to quantify benefits to ecological resources stemming from the Clean Air Act continues to prove problematic, and more work will be needed to bring this task to completion (USEPA 2005). The question posed in this case (Case Study 5) is simple: "What are the benefits, ecologically, of the Clean Air Act?" In a recent endeavor to address this question, a subcommittee of the USEPA's Science Advisory Board (SAB) detailed important elements that need to be present for a scientifically robust benefits analysis to result. The subcommittee found, as we noted already, that valuation "depends." It depends on numerous

Synthesis, Recommendations, and Conclusions

factors including the needs, wants, and desires of the local populace potentially impacted by the decision, as well as the regulatory context and experiences of those in the decisional process. Just as importantly, the SAB subcommittee suggested that one way to drill down on the question and examine the underlying scientific basis is to make sure there are sufficient economic and ecological-environmental data on which to base the analyses. Where the data are lacking, or where there is not a clear connection between the act taken (say, reduction in nitrogen emissions to a nearby bay system) and quantifiable ecological benefits (say, improved fishery for shellfish), then the development of a method(s) to determine the benefits and costs may be difficult, and at worst misleading, in terms of application to other problems. At this juncture it is not clear whether valuation will therefore be restricted to those environmental decisions or ecological risk assessments where there are "sufficient" data, however "sufficient" is defined. It would not seem to be useful if the inclusion of valuation is restricted to a small subset of assessments having rich data sets. Yet to be meaningful, should this in fact be the case? There is as yet no forthcoming answer to this question.

There are different outlooks on the proper role of human values in decision making. As alluded to above, certain public (human) values are incorporated into legislation through representative democratic processes. Therefore, the goals embodied in policies or regulations may be legally fixed. In other cases, however, processes to develop policy or regulation may invite, or require, public participation or outreach. There is a strong role for the public in decisions on the cleanup of Superfund sites in the United States, as mandated by statute and regulations (Blackburn 1994). Requirements may also exist to determine the net social benefit of a proposed public decision, through benefit–cost analysis (BCA), before it can go forward. Currently we are unaware of any comprehensive listing of data that would provide a full evaluation of BCA on a national scale, especially with respect to ecological or environmental benefits.

In private or business decision making, if legal compliance is seen as an acceptable minimum, then no further determination of values is required. On the other hand, if public trust and the local or global stewardship of resources are the ideal, then determination of social values may be needed. But the valuation processes used can be selected at the discretion of the business and usually will emphasize understanding the particular values of various social groups, rather than measuring net social benefits. The case study on mining in Madagascar (Case Study 3) illustrates this issue quite well. That is not to say that the only useful sets of values are those that are linked to consumer wants and needs, but moreover, the sets of values may be much more limited than if the concern was the public at large and any accompanying social improvements.

Today there are a variety of frameworks for environmental decision making, most developed for different reasons or audiences or to satisfy a specific legal requirement. Over time, the role that these frameworks have assigned to stakeholders has increased from that of passive recipient of information to that of active participant in the discussion of goals. Furthermore, the decision-making process increasingly is described as adaptive or reiterative as new information is gained, creating additional opportunities for value-informed discussion. Chapter 2 of this book treats this topic extensively, as it is an important issue related to the use of valuation in the decisional process. Nonetheless, the processes that can be used are highly varied. Therefore, while it is possible to define key principles for the appropriate elicitation and integration of human values, it is not easy to determine the best approach that will satisfy all possible cases.

Conclusion Statement 2. The measurement of values for environmental goods and services should not be based solely on monetary approaches. However, scientists and decision makers should note that the absence of a monetary basis continues to be problematic when attempting to illustrate the importance of ecological risk management decisions in the context of management of risks to humans. And, the absence of a monetary basis does not mean the value is any more or less important than one which can be monetized.

Valuation of environmental goods and services does not always mean the determination of their monetary value (termed "monetization"). This was pointed out several times in this book (Chapters 3 and 4), and will continue to be an important consideration in any future attempts to value environmental goods and benefits resulting from management actions.

As we saw in earlier chapters of this book, there are other ways of determining, and sometimes quantifying, the values that are important for making environmental or risk-based decisions. In Chapters 2 and 3, while economic values and sociocultural values in themselves are not distinct, different disciplines bring different understandings to the study of value. Thus, valuation is not the sole purview of economists, and they should not attempt to undertake valuation in the absence of input from ecologists, risk assessors, and others. The reverse, much as it may trouble ecologists and risk assessors, is also true. These are fundamental points underlying the workshop and this book. Without this input and exchange, the valuation may be sterile and lack the understanding that may come from discussions among scientists and the public.

Scientists working from a sociocultural disciplinary perspective define value as that which is held to be important or valuable within a sociocultural milieu. These scientists tend to be concerned about the social context of value and the motives behind values that are observed or expressed. They are interested in describing the factors that determine a group's well-being. They tend not to approach the issue in the same manner as economists, and for that reason, it is important that the 2 groups collaborate closely (and often) on the valuation process. The work of one will complement and enhance the work of the other. The work of these scientists is mostly qualitative, and they generally avoid abstraction or prediction. As we've discussed previously, typically individuals are more certain of how they value their own well-being and that of other humans than they are of "the environment" or ecological receptors. Unfortunately, this latter point adds yet one more layer of complexity on an already complex issue.

Economists tend to define value as that which individuals are willing to give up in exchange for something else (see Chapters 2 and 3). Their methods are quantitative and therefore lend themselves readily to abstraction, modeling, and prediction (Chapter 4). Economists are comfortable with defining individuals' preferences but not

Synthesis, Recommendations, and Conclusions 169

necessarily the well-being of groups (i.e., social welfare). They can help identify management alternatives that have the potential to increase the well-being of all individuals (i.e., those who would produce welfare gains sufficient to offset losses). However, if an alternative creates winners and losers, while they can help to quantify those gains and losses they are not willing to state whether it is, on the whole, a good choice for society. Few would desire to be the individual faced with communicating to the "losers" in such instances. It is likely that some of these decisions will have a political basis, and there are few elected officials who relish the opportunity to inform their constituents they are the "losers" in any particular decision. As readers may note, we are silent on the issue of how wildlife, fish, or plants would respond (if they only could) as to what they would be willing to give up for something else.

Similarly, ascertaining values from the public on many issues, whether environmental, medical, or educational, can be accomplished readily through a number of approaches (Chapter 2), but it is exceedingly difficult to consolidate those values into one or two overarching values that will satisfy the entire set of individuals. In other words, gaining the qualitative basis for valuation from the public is not the issue, but rather the issue is trying to distill those numerous wants, needs, and desires into one or two that can be managed effectively in the context of the decision being made. The current workshop did not attempt to address this point given all the other issues that needed to be considered over the course of our limited number of workshop days. Some useful advice on this point, however, is given in Chapter 2, where various approaches are described for obtaining information from the public at large. How to then collate, distill, and act on that information is the most difficult aspect of this issue. This may be a topic worthy of a short, focused workshop sponsored by SETAC or another organization in the future.

One of the main reasons that valuation may not have been included specifically in some environmental decisional processes in the years is that values and preferences are seldom consistent among groups of people. Some of the tools available today for ascertaining these values and preferences are not designed to manage the dynamics of individual or group choices, particularly when those choices can alter with changes in economic conditions, or as a result of other factors that may not be fully known to those making the assessments. Often these differences among people and groups of people are reflected through high variability in data that are collected. The differences can further complicate decision making because the decision maker ultimately has to decide which one among the many is the most important for the decision. Nevertheless, this diversity can be helpful in revealing the breadth of underlying social issues. For example, those who are living near an environmental resource, and are most affected by decisions, may have different values (about their quality of life in the proximity of the resource and their willingness to maintain it) from those living farther away. Furthermore, there is a general lack of data about values; in most cases, this lack will be more profound than the lack of ecological data. A more current example is how human values are in flux with respect to the price of gasoline, particularly since the price has risen and fallen substantially after Hurricanes Katrina and Rita. Is the driving public in the United States willing to give up the large, less efficient SUVs so that the demand on gasoline supplies declines and thus drives down the price? The answer to this question will be obtained over the next few years as the demand on oil and petroleum products rises with the ever rising economies of the developing world.

Conclusion Statement 3. Conducting valuations in complex, dynamic ecological-socioeconomic systems remains an area in need of further research and dialog among ecologists, economists, and decision makers. This topic may be suitable for a subsequent workshop sponsored by SETAC or another organization.

This is a key point of the workshop and probably not a surprise to those who have attempted valuation in various situations. There is much more that we have to do before we'll be able to fully integrate valuation into ecological or environmental risk assessment and risk management. One major issue is how to do this in a way that is simple and can be communicated to the public. One might ask, "What is a 'dynamic ecological-socioeconomic system'?" A relatively simple example might be the Everglades National Park (ENP) in south Florida. There are multiple points where valuation and risk management intersect at the ENP: local communities that depend on tourism; tourists who visit the park for purposes of recreation; decision makers at the local, state, and federal levels charged with maintaining the ENP and its environs; as well as the plants, animals, and surrounding habitats that make up the ENP. How should, or is it important for, the decision maker to sort these intersections and determine which ones might be the most important to the decision he or she is facing? Are the dynamics involved so complicated that it is infeasible for the decision maker to even consider them before making the decision? The answer to these questions is not readily apparent and will likely be constructed only after considerable dialog and debate among the various groups, risk assessors, economists, and legal experts. As we alluded earlier, gaining experience may be the only way to develop a useful framework or process that specifically includes valuation methods. In other words, it is important that we first collect case examples where valuation has been done successfully (and satisfactorily) and then assemble the key steps or elements that led to the successful outcome. The current workshop did not afford the time to do this effectively, but the case studies provided in the book should be a good starting point for others to build upon the work we have completed thus far.

The above illustrates, perhaps too simply, that decision making takes place in circumstances of ecological and socioeconomic complexity, and it is not likely to become any easier with time. Lack of understanding of this complexity can hamper cross-disciplinary integration among scientists, economists, resource managers, and elected officials. It is far too easy to remain in our academic-technical comfort zones, seldom to venture forth other than to complain about how little understanding is held by the "other guy." The critical features of complex systems often are difficult for analysts to specify and for decision makers to take into account. In the case of the ENP, tourism is held to be very important because it can be quantified in financial terms. Yet, there is no question that without the plants, animals, and habitats that make up the ENP, there would be no tourism. Another challenge is determining how one can evaluate the significance of complex ecosystems that, by definition, are subject to changes that are sudden or irreversible. At the ENP, natural processes will force change in the ecosystem in ways that may matter to people, particularly when the plants or animals or habitats they traveled to see are no longer there. These aspects complicate the determination of the values people hold, or the trade-offs they would be willing to make. Ecosystems are not static, and therefore the value that people have for them can be expected to change as the systems themselves change. Does this mean that valuations will have some "time limit" before they need to be revised? This question is one that should be addressed by another group should they choose to move beyond the current workshop.

In addition to complexities, ecological uncertainties, including aspects that are naturally variable and aspects that are not understood, are difficult to recognize and to convey. Like ecological systems, values and preferences are also complex; they also have intricate linkages and vary over time and geographic space.

7.2 RECOMMENDATIONS AND RESEARCH NEEDS

A number of additional recommendations and research needs were identified by the workshop participants and are summarized in the following list:

- Environmental management processes should be determined by values relevant to the decision at hand and by the context of the decision.
- The use of analytic frameworks such as the data quality objectives process can help to insure that value-based decisions are supported by the appropriate data.
- Integrated analysis approaches that recognize interdependencies of human and natural systems are less prone to system-specific errors.
- The insights of economists and other social scientists are relevant for all stages of ecological risk assessment, including the identification of assessment endpoints and the design of conceptual models. Therefore, integration should begin with the planning and problem formulation stages.
- Because of its history of collaboration between natural scientists and economists, the field of natural resource damage assessment can provide a useful model for improving integrated assessment.
- Decision makers should strive for "open" decisional frameworks, which allow for and present to decision makers a plurality of values and assumptions and provide opportunities for learning, rather than "closed" processes that lead to a single, "best" outcome.
- Analysts should recognize that it is important to capsulize qualitative findings and try to display them on a par with quantitative findings (and to display nonmonetary with monetary findings) so that they are not ignored in the decision making process.
- Decision makers should likewise recognize that not all of the critical features of a decision can be quantified or monetized; they should take time to understand qualitative findings.
- In-depth comparative analysis is needed of the ways in which governments, NGOs, and the private sector have collected data on sociocultural values and incorporated them into decision making.
- Development of economic and sociocultural tools tailored to the needs of the private sector may enhance the inclusion of public values in private decisions.

7.3 MOVING FORWARD

The SETAC-sponsored workshop on "Valuation of Ecological Resources: Integration of Ecological Risk Assessment and Socio-economics to Support Environmental Decisions," held in October 2003, Pensacola, Florida, provided a useful and necessary step in the evolution and refinement of ecological risk assessment. Just as the early evaluation of potential risk was based heavily on the hazard quotient, today the process is evolving to use, where possible, more robust techniques such as error propagation and/or probability-based estimates. For valuation, the next refinement should be the explicit inclusion of valuation methods into the ecological risk assessment in a way that captures both the qualitative and quantitative values that people place on environmental issues. We encourage our colleagues in SETAC and other scientific groups to use this workshop and its findings to continue the debate and work to further improve the science of ecological risk assessment.

REFERENCES

Blackburn JB. 1994. Ethics, science and environmental decision-making. Environ Toxic & Chem 13:679–681.
[USEPA] US Environmental Protection Agency. 1995. Ecological risk: a primer for risk managers. EPA 734-R-95-001. Washington (DC): USEPA Office of Prevention, Pesticides and Toxic Substances.
[USEPA] US Environmental Protection Agency. 1997. Priorities for ecological protection: an initial list and discussion document for EPA. EPA/600/S-97-002. Washington (DC): USEPA, Office of Research and Development.
[USEPA] US Environmental Protection Agency. 2005. Advisory on plans for ecological effects analysis in the analytical plan for EPA's second prospective analysis — benefits and costs of the Clean Air Act, 1990–2020. USEPA Science Advisory Board. EPA-COUNCIL-ADV-05-001, June 23. Washington (DC): USEPA.

Case Study 1: National Park Establishment, Philippines

Doug Reagan

CONTENTS

C1.1 "Scope" or Type of Problem Category: Resource Management.................. 173
C1.2 Resource Management Goals and Objectives
 for the Valuation Exercise.. 174
C1.3 Stakeholder Identification 174
C1.4 Conceptual Model.. 174
C1.5 Values Identification 175
C1.6 Data Requirements... 176
C1.7 Analysis Plan 177
C1.8 Characterization and Communication 177
C1.9 Did the Outcome Meet Expectations? What Were the Lessons
 Learned?... 177

C1.1 "SCOPE" OR TYPE OF PROBLEM CATEGORY: RESOURCE MANAGEMENT

When the US Navy's lease at Subic Bay, Philippines, ended in 1992, the Philippine government designated the area and adjacent lands as the Subic Bay Freeport Zone (SBFZ). Land management by the navy had maintained several thousand hectares of mature tropical rain forest for purposes of survival training. The forest and portions of the surrounding area were inhabited by indigenous people, the Aeta, who were seeking legal title to nearly a third of the remaining forest as a traditional land claim.

Much of the area within the boundaries of the SBFZ had been developed by the US Navy for military operations, and virtually all of the area within the zone but outside of former navy jurisdiction had been developed for residential, commercial, or agricultural uses. Mangroves elsewhere in Subic Bay had been converted to rice growing or other uses, and freshwater stream discharges to the bay from areas outside the base had impaired water quality as the result of sedimentation from cleared areas or inputs of raw sewage from urbanized development.

Most of the SBFZ had been impacted by deposition of +20 cm of ash from the eruption of Mount Pinatubo, causing the loss of sensitive forest plants, burying coral reefs, and providing additional substrate for the expansion of seagrass beds. Portions of the area were difficult to access, and some presented safety hazards that restricted or hampered assessments. These included the presence of National Peoples Army (NPA) guerillas in remote forest areas, smugglers and occasionally pirates in outer coastal areas, and the more usual forest hazards such as cobras.

In 1999, the Ecology Center of the Subic Bay Metropolitan Authority (SBMA, the Philippine government authority) initiated the development of a Protected Area Management Plan to provide the framework for establishing and managing the ecological resources as a national (natural) park under the National Integrated Protected Area System (NIPAS) Act. All work was to be performed with regular involvement of stakeholders, and all planning activities were to be completed within a 24-month time frame.

C1.2 RESOURCE MANAGEMENT GOALS AND OBJECTIVES FOR THE VALUATION EXERCISE

The overall goal of this project was to develop a Protected Area Management Plan for establishing a new national park at Subic Bay, Philippines. The objectives were as follows:

- Conduct natural resource inventories.
- Delineate protected area (park) and buffer zones.
- Develop forest and coastal management plans.
- Establish adequate institutions for monitoring and enforcement.
- Develop alternative livelihood strategies for affected groups.
- Produce a final Protected Area Management Plan for park establishment.

C1.3 STAKEHOLDER IDENTIFICATION

Who were the primary, secondary, and tertiary stakeholders?

- Primary: Subic Bay Metropolitan Authority (Philippine government) and Aeta (indigenous people)
- Secondary: commercial interests, local government units (LGUs), residents, and nongovernmental organizations (NGOs; local and international)
- Tertiary: future generations that should not be denied future opportunities by decisions made in this generation

C1.4 CONCEPTUAL MODEL

Ecological service flows among natural resources and from natural resources to humans (e.g., food, construction materials, and medicines) were modeled to ensure that major services were included in ecological, social, and economic evaluations. Conceptual exposure models in the context of risk assessments were not employed. Human services were evaluated by economists to provide a cost–benefit analysis for planning mitigation (e.g., alternative livelihood strategies, and recreational opportunities) and to develop and understand the economic implications of various options for park establishment.

C1.5 VALUES IDENTIFICATION

A process for identifying ecological values and human use values of natural resources of the study area identified ecological, economic, and social values with stakeholder involvement as a basis for scoping data collection, drawing boundaries, and developing a detailed management plan. The premise of this framework was that there are ecological values (i.e., properties) that are essential for defining healthy ecosystems, and that some components or functions of the ecosystems are used by humans (e.g., fruits used for food, and wood and bamboo used for construction). This framework for values identification included a hierarchical model of ecosystem values (e.g., healthy sustainable ecosystem, biodiversity, and ecosystem processes) coupled with a concurrent process for identifying societal values, including economic values.

Presentation materials were developed to illustrate the services provided by physical resources to biological resources, by biological resources to other biological resources, and by the environment to people. One such illustration depicted the importance of the natural ecosystem for providing food, medicines, ritualistic materials, and other needs of the indigenous Aeta (Figure C1.1). This graphically illustrates the importance of the forest in maintaining the well-being and cultural identity of the Aeta and demonstrated this relationship to all stakeholders.

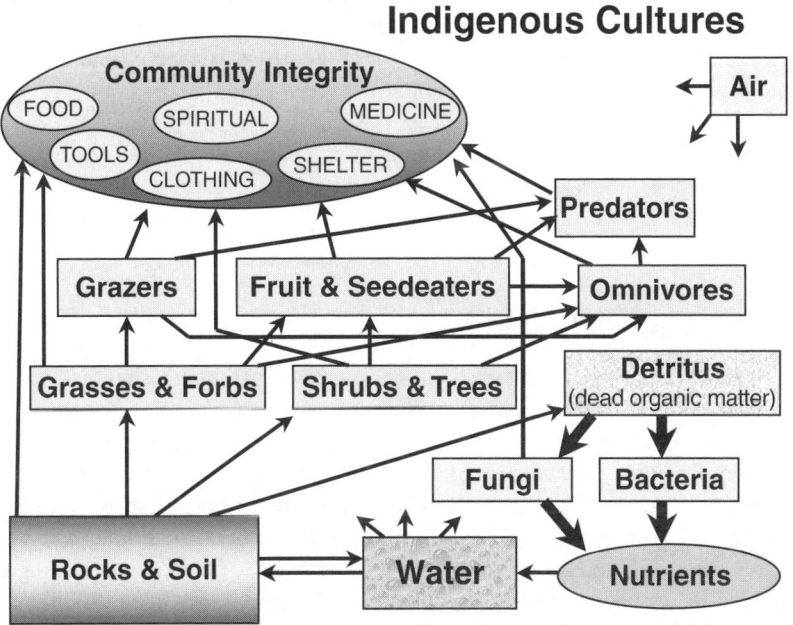

FIGURE C1.1 Environmental service flows for indigenous communities.

The framework for values identification was presented at the project initiation workshop, and stakeholders were asked to list their concerns and values for park establishment. The team developing the management plan used this input to develop the plan and promised to attempt to address each of the identified values. A predictable outcome of the process was that some values identified by different stakeholders were in conflict (e.g., endangered species protection and timber harvesting).

For planning purposes, properties of ecosystems were treated as ecological values corresponding to societal values, including the economic value that could be estimated for some natural resources. Planning for values that were in conflict was addressed spatially or temporally (e.g., different uses in different areas, and some areas off-limits at some seasons to preserve important functions such as providing a breeding habitat for fish).

In some instances, the values identification process linked economic, social, and ecological values of some species. For example, the analysis identified the ecological value of the giant fruit bats, which maintained a large roost in the forest, as pollinators and seed dispersers. The societal value of the bats was as food for local communities and indigenous inhabitants, and the bats provided an additional economic benefit as a tourist attraction.

Affected resources or their uses were addressed by mitigation planning including the development of alternative livelihood strategies for communities and individuals adversely affected by park establishment. Planning included proposals for training and hiring fishers affected by fishing limitations to support ecotourism and other industries related to park establishment.

Park boundary establishment, designation of land use areas within the park, and buffer zones addressed different spatial scales. Identification and planning for resource protection in buffer zones employed a watershed approach rather than concentric zones, and subdivisions within the park for core area (research and indigenous use), sustainable use, multiple use, habitat use, restoration, and existing economic use zones.

Temporal planning was tiered. Detailed management plans were developed for each of the first 5 years (establishment and implementation). Overall planning was for 50 years, with periodic monitoring and review for adaptive management throughout.

C1.6 DATA REQUIREMENTS

Data on natural resources and uses were necessary to inform decision making. Data were collected on physical resources (e.g., water quality, physical oceanography, and geology and soils), biological resources (e.g., mangroves, stream biota, coral reef structure and composition, forest flora, and terrestrial fauna), and social issues (e.g., demography, fisheries, recreational uses, land use, and forest resource survey). Resource conditions and external influences (e.g., continued effects of the eruption of Mount Pinatubo) were also noted for each of the resources as a basis for determining park boundaries, designation of use zones, and areas of potential resource restoration.

Available information was obtained and reviewed to evaluate its usability to support decision making for park establishment and management planning. Where data

Case Study 1: National Park Establishment, Philippines 177

gaps were identified, additional data of the appropriate type and quality were used to fill the identified data gap. Only data relevant to the decision at hand (e.g., to draw boundaries and develop the various plans) were collected. For example, we did not collect data on insect biodiversity, although we recognized that such information would be important for future management. Data on distribution of resources (e.g., mangrove areas, coral reefs, and fruit bat roosting areas) were compiled in GIS format for presentation, planning, and plan implementation purposes.

C1.7 ANALYSIS PLAN

Once data collection was complete, results were presented at a stakeholder workshop to identify additional possible sources of information and augment the database. We then presented a general description of how the data would be used and invited additional input on issues and constraints. This approach identified several important issues, not previously recognized, and because of the transparency of the process substantially reduced concerns with park establishment early in the project. Because stakeholders were involved from the outset, the process became increasingly participatory. Documentation of inputs and how they were used was important in building trust and support for the park establishment in spite of the inherent conflicts between preservation and development interests.

Because data were collected in a decision-making context that was developed on the basis of identified ecological and human use (social and economic) values, a number of disciplines were required to develop the park plan. Ecologists, hydrologists, oceanographers, geologists, anthropologists, economists, and community involvement specialists continued to work with planners to develop a detailed draft management plan. Pertinent data from field surveys, aerial photography, land use planning agencies, and other sources were compiled in the GIS database for plan development.

C1.8 CHARACTERIZATION AND COMMUNICATION

Information was analyzed by experts and presented to stakeholders at a series of 5 2-day workshops distributed throughout the planning process for their review and as the basis for obtaining and using additional input. Copies of the 8-volume draft and final plans were available for public review and comment. Findings were conveyed to decision makers in workshops, briefings, and written draft and final plans, and news releases on plan progress were prepared and distributed at various project milestones. Project staff were also invited to present and discuss the project in various forums including schools, chamber of commerce meetings, Rotary Club, and the like.

C1.9 DID THE OUTCOME MEET EXPECTATIONS? WHAT WERE THE LESSONS LEARNED?

The outcome met and in some areas exceeded expectations. Park establishment was endorsed by all stakeholder groups, and the Philippine government (SBMA) formally adopted the plan. Subsequent plan implementation has been hampered by political uncertainties and funding constraints, but it continues to move forward with broad stakeholder support. The plan provided discrete elements in need of funding, and some of these have already been addressed by outside institutions.

While the plan was in the final stages of completion, the Aeta (indigenous people) obtained an ancestral lands claim for portions of the proposed park and supported the park management plan because it integrated their concerns and provided a comprehensive context for management and protection of the resources they valued. Commercial interests accepted the plan because it enhanced regional quality values, protected economically important resources, and provided opportunities for new recreation-based development as part of the plan. Buffer zone planning was done on a watershed rather than concentric zone basis and created a regional framework for maintaining water quality in all of Subic Bay. As such, it was a starting point for addressing regional issues and supporting the goals of local communities and government units.

Ongoing communication and community involvement were critical to the plan's success. Each workshop was designed to present results, discuss the current stage of planning and how inputs from previous workshops had been used to develop or modify the plan, remind people of the overall value-based process, and provide additional opportunities for active input. At the 4th of 5 workshops, participants were asked to develop a vision statement for the park. This exercise brought together representatives of NGOs, developers, commercial interests, local governments, indigenous people, and others, each of whom had different emphases but acknowledged the broader needs. Each stakeholder group identified the need to sustain the values from healthy ecosystems, protect key resource areas (e.g., endangered species, critical habitats, and nursery areas for fish), manage the sustainable utilization of renewable resources (e.g., fish and bamboo), permit controlled development, and acknowledge the rights of the indigenous Aeta. At this juncture in the planning process, stakeholder involvement shifted from participation to project ownership, ensuring broad-based long-term support for park establishment and operation.

Key elements of project success were as follows:

- Appropriate project organization and planning
- Comprehensive and systematic framework for values identification
- Participatory, transparent, and informative planning process
- Provision for stakeholder empowerment in park management
- Development of alternative livelihood strategies for affected stakeholders
- Scientifically and technically based planning approach

Case Study 2: Large PCB-Contaminated River under Superfund

Interface between Remediation and Restoration

Katherine von Stackelberg

CONTENTS

C2.1 Scope or Type of Problem 179
C2.2 Resource Management Goals and Objective(s) for the Valuation Exercise 180
C2.3 Stakeholder Identification 181
C2.4 Conceptual Model 181
C2.5 Values Identification (~Assessment Endpoint) 182
C2.6 Data Requirements 182
C2.7 Analysis Plan 182
C2.8 Characterization and Communication 184
C2.9 Did the Outcome Meet Expectations? If Not, What Were the Lessons Learned? 185
C2.10 Additional Reading 185

C2.1 SCOPE OR TYPE OF PROBLEM

The site is a large polychlorinated biphenyl (PCB)–contaminated river system that is partially freshwater and partially estuarine. From the 1930s to the late 1970s, PCBs were manufactured and marketed under the trade name Aroclor for use in dielectric fluids, hydraulic fluids, solvent extenders, flame retardants, organic diluents, inks, dyes, paints, and adhesives. PCBs are persistent in the environment and not very soluble in water. The oil-like PCBs were considered an ideal insulating fluid quickly replacing substances that easily flammable, and they also were used in carbonless copy paper, newsprint, and caulking compounds.

The basic chemical structure that all PCBs share is the biphenyl molecule. All PCBs have at least 1 chlorine atom attached, and there are 10 possible locations on the biphenyl molecule where chlorine atoms can attach. There are 209 individual PCB congeners.

Manufacture of PCBs leads to the production of mixtures of the different types of PCBs, and the trade name Aroclor represents such a mixture. The mixtures produced varied in their weight percentages of chlorine, which was indicated by the last 2 digits of the 4-digit number following the Aroclor name (e.g., Aroclor-1260 contains 60% chlorine by weight). The most commonly produced Aroclors were 1016, 1221, 1242, 1248, 1254, and 1260. The exact proportion and identity of each of the PCBs in the mixture were not standard (even for a given manufacturer between lots) or relevant since the properties of the mixture depended on the total weight percentage of chlorine, not the allocation of congeners that achieved the total weight.

PCB mixtures degrade once they are released into the environment, and in particular they slowly become dechlorinated. Consequently, the nonstandard mixtures continually change, which makes assumptions about the components of the mixture tenuous at best. All the toxicological studies upon which ecological toxicity reference values or human toxicity values are based were conducted on laboratory-derived Aroclors rather than the mixture of PCBs found in the environment.

The United States Environmental Protection Agency (USEPA) conducted a survey in the mid-1970s showing elevated levels of PCBs in fish. The state environmental agency subsequently confirmed those findings, and yearly monitoring of fish has been conducted since that time. In the late 1970s, the state Department of Environmental Conservation and Department of Health banned all fishing in the freshwater portion of the river and closed all commercial fisheries. In that same year, a massive flood (of the magnitude that only 1 such flood is expected to occur every 100 years) caused a large movement of contaminated sediments from the upper portion of the river into the lower river.

The site is a Superfund site being managed by the USEPA and underwent a Remedial Investigation/Feasibility Study (RI/FS) to evaluate potential remedial alternatives. The site is also the subject of a natural resource damage assessment (NRDA) claim by trustees such as the Fish and Wildlife Service (FWS) and the National Oceanographic and Atmospheric Administration (NOAA) in order to pay for restoration of natural resources.

C2.2 RESOURCE MANAGEMENT GOALS AND OBJECTIVE(S) FOR THE VALUATION EXERCISE

Comprehensive Environmental Response Compensation and Liability Act (CERCLA) remediation: when will predicted risks and hazards achieve acceptable levels under various remedial alternatives (including no action, monitored natural attenuation, and a number of active remediation scenarios such as capping, dredging, and in situ bioremediation)?

NRDA: the objective of the NRDA process is to compensate the public, through environmental restoration, for losses to natural resources that have been caused by releases of PCBs into the environment. Thus, the goal is settlement of the NRDA claim by quantifying and monetizing injuries to impacted trust resources, including fish, birds, mammals, and their supporting habitat.

Case Study 2: Large PCB-Contaminated River under Superfund

However, the final assessment of natural resource damages does depend, to some extent, on the results of the RI/FS process because the potential for restoration and the nature and extent of future damages will differ depending on the extent of PCB cleanup and/or removal under the CERCLA remediation plan. Following a CERCLA remediation decision, the trustees developed a detailed restoration plan for public comment that provides a description of the restoration measures, including descriptions of the specific projects that will be undertaken to restore, rehabilitate, replace, or acquire natural resources and thereby compensate the public for harm caused by PCBs.

C2.3 STAKEHOLDER IDENTIFICATION

The list of stakeholders at this site includes the following:

- The potentially responsible party (PRP), who has a team of scientists and lawyers working on this case, and a visible public relations campaign to counter proposed remediation and restoration activities
- The sport fishing community (longtime residents of the area who use the river recreationally)
- A small subsistence fishing community (primarily Vietnamese, who speak little or no English, and Native American populations)
- The Department of Environmental Conservation
- A number of grassroots organizations
- The electronic and print media, which are the public's primary source of information about research and regulatory activities regarding the site
- The USEPA
- Federal trustees, including FWS and NOAA

Each of these stakeholders has some competing and some overlapping interests. The list of stakeholders is the same between CERCLA and NRDA, although the priority of the stakeholder interests might differ between the 2 programs.

C2.4 CONCEPTUAL MODEL

The conceptual model for the site, a multi-product electrical manufacturing facility, begins with the identification of sources of PCBs to the system. Ongoing releases, as well as the sediments in the river, comprise the source of PCBs. Resuspension of highly contaminated sediments from the most highly contaminated area is implicated in the continuing downstream migration of PCBs. In addition, PCB releases from the former plant through the migration of PCB oil through bedrock have also occurred, although the extent and magnitude of this release continue to be studied.

PCBs enter the aquatic food web from contaminated surface water and sediment via uptake by phytoplankton, water column invertebrates, and benthic invertebrates. Forage fish (e.g., smaller fish that consume primarily invertebrates) are exposed to PCBs by consuming invertebrates, and in turn serve as a PCB source to predator fish.

Similarly, PCBs that enter the food chain via benthic invertebrates are transferred to bottom-feeding fish.

Piscivorous birds and mammals, and people, are exposed to PCBs that accumulate in fish and biomagnify in the aquatic food web. Organisms that consume birds and mammals that have consumed fish face potentially higher exposures. Waterfowl such as mallards also contain elevated PCB concentrations as a result of exposure to sediment, surface water, phytoplankton, and zooplankton contaminated with PCBs.

C2.5 VALUES IDENTIFICATION (~ASSESSMENT ENDPOINT)

The goals are the protection and sustainability of populations of fish and wildlife using the river.

The assessment endpoint under the NRDA is similar but expressed in different terms: the quantity of injured resources (e.g., the percentage of loss of resource services). In addition, the NRDA evaluates the potential for recreational fishing losses. The NRDA plan is designed to provide ecological and social benefits. For example, a key element of the approach is ensuring that the restoration addresses the full geographic and ecological scope of injuries to natural resources. In this case, the preferred restoration alternative includes wetland preservation, wetland restoration, and reduction of nonpoint source runoff loads into the bay from cropland through conservation tillage and installation of vegetated buffer strips along streams.

C2.6 DATA REQUIREMENTS

CERCLA remediation: requires information on movement of chemicals through the environment. Further, the potential effects of the chemicals are to be assessed with respect to on human and ecological receptors.

C2.7 ANALYSIS PLAN

CERCLA remediation: Under this regulatory program, the analysis required a full risk assessment. The risk assessment model was used to determine the potential for risk and hazard. Analysts used a probabilistic bioaccumulation model to develop predictions of contaminant concentrations across trophic levels in the aquatic food web. Predicted fish body burdens served as inputs to the ecological dose models. In addition, there were a number of population studies (e.g., census studies) and toxicity studies to evaluate conditions in the field. A bird study was conducted, which measured concentrations of PCBs in eggs and hatchlings at various locations together with behavioral metrics. A statistical model was used to evaluate the relationship between PCB concentrations and behavioral outcomes, and differences in concentrations and outcomes between contaminated and reference areas.

An egg injection study was used to determine responses of fish from the river to injected concentrations of PCBs. Finally, a mink feeding study, using fish from the river, was used to develop a dose–response relationship between PCB exposure concentrations and reproductive outcomes.

Case Study 2: Large PCB-Contaminated River under Superfund

NRDA: Injury determination typically falls under one of 2 categories. The first category establishes injury based on exceedances of regulatory criteria, such as water quality standards. This may include the violation of established standards, the existence of state health advisories warning against the consumption of contaminated biota, and closures or restricted use of resources. The second category establishes injury based on physical, chemical, or biological changes in the resource resulting from contaminant exposure. Examples of these injuries include changes in an organism's physical development, health, reproductive success, or behavior. Ideally, injury to the resource is quantified in terms of the loss of services that the injured resource would have provided had the contaminant release not occurred. Loss of services can also include impairment of habitat that a resource provides or diminished human use of a resource. These kinds of studies are always conducted in terms of comparisons between results in the contaminated area and those in an appropriate reference location or locations. A number of these studies were conducted to determine injuries to recreational fishing resources, birds, and mammals along the river.

Restoration scaling is used to determine the amount of the preferred restoration alternative that is required to compensate the public for injuries to natural resources. This analysis used a total value equivalency (TVE) study to scale restoration actions under the preferred alternative. The study explicitly elicits public input regarding their priorities and values for restoration alternatives, which are presented as scenarios. This ensures that there is public input on the selection of alternatives.

To obtain public preferences and values, a survey was conducted with residents of 10 counties surrounding a 40-mile stretch of the site. The survey focused on 4 types of natural resource restoration programs for the area. The scenarios were designed by assessing relevant options and responses from respondents in survey focus groups and pretests. The scenarios included different levels of the following attributes:

1. *Restore wetlands.* Wetlands restoration provides increased spawning and nursery habitat and increased food for a wide variety of fish, birds, and other wildlife. This provides wildlife services similar to, but not the same as, those injured by PCBs. Priorities and values for restoration of wetlands can also be applied as indicators of the priorities and values for other habitat enhancement projects. Restoration levels range from taking no action up to a 20% increase in wetlands within 5 miles of the site.
2. *Reduce runoff.* Controlling runoff improves water quality by lessening algae growth and improving water clarity. This improves aquatic vegetation and habitat for fish and some birds, and improves recreation. The runoff control in this case provides similar, but not the same, services as those injured by PCBs. The runoff control levels considered range from no change in the amount of runoff up to a 50% reduction, reflected by changes in water quality measures.
3. *Enhance outdoor recreation.* Enhanced recreation includes increasing facilities at existing parks, such as adding picnic grounds, boat ramps, and biking and hiking trails; and developing new parks. The levels of recreation enhancements considered range from no improvements up to a

10% increase in facilities at existing parks and a 10% increase in new park acreage.
4. *Remediate PCBs in the sediments of the assessment area.* Removing PCBs will reduce the number of years until the injuries to wildlife are eliminated. The levels of removal considered result in the number of years until PCBs are at safe levels (i.e., a return to baseline conditions) ranging from 100 years (no additional removal) to 20 years (intensive remediation).

The TVE study supports restoration planning by providing a large-scale perspective of public preferences across alternative types of restoration programs, and providing a method to scale programs that provide equivalent value to the service flow losses. However, the study does not provide a selection of individual projects such as specific wetland acres or specific recreational facilities.

The survey describes each of the 4 natural resource restoration programs and asks a variety of questions to elicit preferences about the programs and the program levels. Next, the survey includes 6 stated preference choice questions, where respondents state their preferences by choosing which of 2 alternatives (A or B) they prefer, where each alternative has a specified level for each of the 4 restoration programs.

C2.8 CHARACTERIZATION AND COMMUNICATION

CERCLA remediation: There was an extensive public communication campaign designed for this site. Early on in the process, a Science and Technical Committee (STC) was established that followed and reviewed all proposed analyses, and then reviewed all the results. Numerous public meetings were convened at different locations and times. The public, PRP, and any interested party were invited to provide comments on every report, and a responsiveness summary was issued to address every comment made. The USEPA set up a website and provided all the reports, responsiveness summaries, peer review reports, responses to the peer review reports, and information in the docket as PDFs on the website.

For the NRDA, similar steps were taken to involve the public. The public was invited to comment on the proposed approach for evaluating restoration options, as well as the plans for and the actual technical analyses. In addition, the TVE study asked respondents questions such as how aware they were of each of the 4 natural resource topics presented (wetlands, PCBs, outdoor recreation, and runoff control) before receiving the survey. Respondents reported being moderately to highly aware of the topics, with more than 80% reporting they were somewhat to very aware of each topic. Analysts expect that higher awareness typically enhances the reliability of responses and reduces the burden of communication in survey design.

C2.9 DID THE OUTCOME MEET EXPECTATIONS? IF NOT, WHAT WERE THE LESSONS LEARNED?

The outcomes of the risk assessment and other modeling analyses were sufficient for decision makers to decide on a remedial alternative under CERCLA. However, the remedial decision was made completely independently of the NRDA process, and at no time was there an awareness of the need for the NRDA to have particular analyses or results expressed in particular ways. The NRDA proceeded according to rather detailed regulations under the Code of Federal Regulations (43 CFR 11). In both cases, regulators and analysts made every effort to follow guidelines specific to the regulatory program under which they were operating. In several cases, this led to inefficiencies in analyses and a duplication of effort. Part of that is very simply a function of resources and "who is paying" and therefore responsible for the particular outcome. For example, both the ecological risk assessment and the NRDA were interested in the results of the bird study that evaluated the relationship between PCB concentrations in eggs and hatchlings and behavioral outcomes in contaminated areas versus reference areas. The ecological risk assessment used this information to develop a toxicity reference value for tree swallows. The NRDA used this information to quantify injuries to songbirds. Both regulatory programs had a vested interest in different aspects and interpretations of the results.

C2.10 ADDITIONAL READING

For an additional perspective on the relationship between the NRDA and CERCLA, readers are encouraged to review Barnthouse LW and Stahl RG, Jr. Quantifying natural resource injuries and ecological service reductions: challenges and opportunities, Environmental Management 30(1):1–12.

Case Study 3
Extractive Development and the Valuation of Biodiversity and Community Needs in Madagascar

Elaine Dorward-King

CONTENTS

C3.1 Statement of the Problem 187
C3.2 Resource Management Goals 188
C3.3 Stakeholder Engagement 188
C3.4 Conceptual Model 189
C3.5 Economic Opportunity 190
C3.6 Values Identification 191
C3.7 Significant Outcomes and Lessons 193

C3.1 STATEMENT OF THE PROBLEM

Madagascar is an island nation of extreme poverty. It is located off the east coast of Africa and has 14 million inhabitants on a land area 2.5 times as large as Britain. It is home to the Malagasy people and a large number of unique flora and fauna. A growing human population and destructive subsistence agricultural practices threaten many species.

A Rio Tinto subsidiary, QIT Madagascar Minerals (QMM), is conducting studies to evaluate the mining of dune and sand deposits of ilmenite in southeastern Madagascar. Ilmenite is a titanium-bearing mineral; titanium is used to make pigments for paints and plastics, and is used in the manufacture of steel and metal alloys. The mine and the economic opportunity it will bring may offer the best chance of preserving some of the last remnants of littoral (coastal) forest on Madagascar. Less than 10% of the original littoral forest remains on Madagascar, and much of what remains, present in small pockets along the length of the island, has been significantly degraded by human activity.

The sensitivity of the values associated with biodiversity and social-cultural traditions requires QMM to be extremely careful in how it proceeds in developing the project. It has adopted an integrated approach to advancing the technical aspects of the project while investing in understanding both the local and international ecological and social-cultural issues.

C3.2 RESOURCE MANAGEMENT GOALS

The multiple resource values at the site require management goals to include the efficient and effective extraction of the desirable minerals from the sands, conservation of remnant higher quality littoral forests, restoration of ecological function for other littoral forests that will be destroyed during the course of mining through the dunes, and creation of forests containing fast-growing native and non-native plant species of commercial and domestic value for the local people.

Conservation and restoration of the littoral forest are required to protect valuable habitat for many endemic plant and animal species, including several lemur species that are threatened or in danger of extinction. Restoration of the littoral forests and creation of sustainable commercial forests are needed as well to provide the fuel and food resources essential for the traditions of the local Malagasy people.

C3.3 STAKEHOLDER ENGAGEMENT

Discussions about a possible mineral sand development project in southeastern Madagascar extend back many years. QIT Madagascar Mining has been consulting with the Madagascar government on this project since the mid-1980s.

Early on in the project, several international environmental NGOs leveled significant public criticism at QMM for the project, citing concern for the destruction of the littoral forests that would be removed during mining of the mineral sands. Other interested parties believed that developing the deposit could contribute significantly to both the local and national economies in Madagascar over the medium to long term, if the mining project was part of an overall development plan for the southeastern region of the island.

The lack of established background data made comprehensive studies an essential part of planning from the outset. QMM embarked on a significant program of ecological and social research to evaluate the status of the forests and surrounding areas, and to begin understanding what values should be emphasized in restoration if the mining development was to be seen as sustainable. A team of national and international specialists was commissioned in the late 1980s to undertake baseline studies of the existing natural and social environments. A conceptual mining project was outlined to define the area over which to carry out the studies. The study program was developed through discussion with a number of organizations, including the World Bank, the Canadian International Development Agency, Conservation International, the World Wide Fund for Nature in Madagascar, and other local interest groups. This work was carried out up to 1992 and concluded with a report available to all involved in the process.

Since that time, QMM has had to address key challenges in establishing the project, which represents the largest private investment in Madagascar and the only current economic opportunity in the southeastern region. These challenges include the following:

- How can a mining project be compatible with the values seen in biodiversity?
- How can the project aid in reducing poverty, a necessary condition for reducing pressure on biodiversity?
- How can local community acceptance of the project be obtained, and opportunities for local people enhanced?
- How can operations be set up best, given the lack of social and physical infrastructure to receive a project of this magnitude?
- How could informed support by the local, regional, and national authorities be developed?

Stakeholders in the project include the local Malagasy people, the local and regional governments, the national government, QMM, environmental and social NGOs, and intergovernmental aid agencies involved in the local area. Working with the other stakeholders, QMM has sought to integrate the project into the fabric of the region. The company has taken several key actions:

- Approached the management of the proposed conservation areas through dialogue led by the local authorities with the participation of all interested parties, leading to the adoption of a formal "convention."
- Actively supported a regional planning process that is locally based and driven (and which has identified the mining project as a major positive factor in the development of the area).
- Planned and implemented a consultation process at the local, regional, national, and international levels with attention in preparing local consultations and adapting processes to local circumstances to facilitate maximum understanding of issues and values, while enhancing everyone's ability to express concerns and aspirations. Ongoing consultations help identify and test solutions.

C3.4 CONCEPTUAL MODEL

Remnant littoral forests of special botanical interest and value were identified by QMM's baseline studies. These forests are under pressure from local people using the wood for charcoal fuel and for building. This subsistence activity has had marked detrimental impact, and there is no longer littoral forest on approximately two-thirds of the mineral deposit area. It is generally accepted that the remaining forest will be effectively destroyed within the next 20 to 40 years. The baseline studies showed that conservation measures would be essential to protect species and suggested that a mining project could proceed in parallel with measures to preserve the area's most important environmental features. The studies proposed the creation of zones to conserve flora and fauna, including part of the littoral forest, while allowing other areas of the remaining forest to be gradually cleared over a period of many years.

In these areas, it was suggested, replanting with a commercial forest crop should follow any mining activity. These could provide an alternative source of fuel and construction wood for local people, thereby helping to relieve the present pressures on the forests.

Mined areas will be returned to natural contours, as has been done at similar mineral sand operations in South Africa and Australia, including the Rio Tinto-managed Richards Bay Minerals. QMM is committed to protecting some core forest areas (some 10% of the potential mining area) as permanent conservation areas. Land immediately contiguous to the conservation areas will be rehabilitated in native species, while other disturbed land will be rehabilitated with fast-growing commercial species that will provide an alternative source of firewood, charcoal, and lumber. Much of the land in the mining area is already degraded by overuse. Highly valued wetlands, which are also a source of income to local people through the harvesting of reeds (Mahampy), will all be restored.

In order to be able to address these challenges and commitments, QMM has put into place a permanent social and environmental team of Malagasy professionals; there are more than 150 professionals and labor staff in the project area. The team has carried out research in both the social and biological aspects of the region.

A nursery has been developed to experiment with germinating, growing, and transplanting both commercial and native littoral species that will be used for rehabilitation. Full-scale experiments have been carried out in situ for the past 4 years. A plantation of 500 hectares (ha) of fast-growing species outside the mining area began in 2001 to create an initial source of wood supply to reduce pressure on proposed conservation areas. The experiments also include experimental rehabilitation programs of the local reeds, done with the participation of the women who are the main users of this resource.

Another lasting dimension of the project is the creation of a port facility, required for the export of the mineral, which will be a public facility financed as a public and private partnership, creating new possibilities of access to and from the region and, consequently, other economic opportunities.

QMM submitted an integrated Social and Environmental Impact Assessment to the government of Madagascar in May 2001. The government issued its environmental permit in November 2001 along with extensive requirements, which forms the Environment Management Plan (EMP) and constitutes contractual obligations for QMM.

C3.5 ECONOMIC OPPORTUNITY

The 3 large ilmenite deposits discovered by Rio Tinto in Madagascar cover roughly 6000 ha and have an estimated worth of hundreds of millions of dollars in revenue. These deposits make up approximately 10% of the world's known ilmenite reserves.

Ilmenite is a heavy black mineral that is mixed naturally with the white beach sand of the Madagascar southeastern shore. The ilmenite-containing sands are dredged from large ponds and centrifuged with water in separators, and the black mineral is removed. The white sand is returned to the site, and the dredging equipment moved along the dunes. The project will occupy an area of about 200 meters by 200 meters at a time, and move forward across the site at a rate of about a meter a day.

Case Study 3

In order for the mining project to begin, a port and breakwater will have to be constructed (US$80 million), a $75 million separation plant built, and some roads and other infrastructure built. Under the framework agreement, QMM has the obligation to build and operate the port. The port will be a public port, and will be managed by QMM for both their and the public's use.

QIT has spent about $30 million over the 15 years it has been assessing the deposit. Now that the research on forest rehabilitation is in full swing, that has edged up to about $4 million a year. A decision about whether to go ahead with the project was made in 2005 and construction has begun.

While the Malagasy government will ultimately have to decide how best to promote responsible economic development to contribute to the alleviation of poverty and to conserve natural and social values, there is increasing understanding and consensus among many of the regional stakeholders that the mineral sands mining development presents an important opportunity to reverse the current trajectories of increasing environmental destruction and desperate poverty.

C3.6 VALUES IDENTIFICATION

As discussed above, considerable effort was expended to understand social-cultural values in conjunction with their impacts upon biodiversity and the values associated with that biodiversity. The baseline and ongoing studies have defined the complexity of the system well enough to inform stakeholders and to guide the decision-making process. As QMM and the Madagascar government make decisions about project development in the context of regional development, understanding and elucidating social-cultural values will continue to improve.

Understanding the apparently contrasting values of international environmental organizations as well as the values and near-term needs of the local community was of significant importance to QMM. This was necessary to ensure that the project was planned and developed to maximize opportunities for addressing immediate and midterm local needs while also establishing mechanisms for the long-term sustainability of economic, public health, and environmental improvements and protections that would emerge during the project and mine life.

QMM addressed this issue in several ways, including extensive stakeholder engagement over a long period of time, seeking to continuously inform and have dialogue with interested parties as thinking and understanding evolved. Consultation with the Malagasy occurred at all levels — local, regional, and national — and was highly successful largely because it was very open and inclusive, as well as done over a long period of time. Trust by the local and regional people involved was developed not only by involving them in discussions about plans, issues, and problems, but also by working with them in the field to demonstrate that ideas being proposed as solutions would work and were practical. For example, nursery work and fieldwork, involving local workers, was able to demonstrate that both the native species of plants and those proposed for commercial purposes would grow and prosper in the rehabilitated dune sands. One project, accomplished in close partnership with local women, was to demonstrate that Mahampy could be established in newly created wetlands. The company built a pond and planted reeds, and both QMM and local women monitored their growth. At the appropriate time, reeds were harvested and tested by the local women in the traditional uses to ensure that functionality was not impacted. This experiment, done over more than 1 year, contributed to the trust building as well as to the understanding of the overall restoration program.

In addition, input and involvement from the external scientific and NGO community were actively sought. QMM was very public in its call for advice and expertise to address the many biodiversity issues. While in the beginning there was some skepticism and reluctance to engage with the mining company, this has changed over time. The ecological research area is serving as a field laboratory for both QMM scientists and academic researchers from Europe and North America. Collaboration in order to maximize efforts has led not only to results of interest to academic science, but also to results that may benefit the applied ongoing restoration work.

It is recognized that unless the significant social needs of the people in the area of the project are addressed, ecological restoration successes, either in the mining area or outside, will not be sustained. QMM has placed considerable importance on addressing the social issues. They have found that as daunting as the challenges presented by the natural environment are, those challenges presented by the social-cultural context have been much more difficult to progress. The same open process of engagement has been employed, but the scope and complexity of the issues and the difficulty in prioritizing public needs continue to require a thoughtful and deliberate approach. Indeed, it is impossible to address the local social needs in the absence of a regional economic development plan that involves building capacity and infrastructure over time. QMM has worked closely with the authorities as a Regional Development Plan has been developed, contributing technical knowledge and sharing project alternatives and plans. Key social issues that require attention include education, clean water, adequate nutrition, sanitation, basic health care, and appropriate prevention and treatment of malaria and sexually transmitted diseases (these are very high, although HIV prevalence is currently low throughout Madagascar). Addressing these symptoms of poverty requires that people have opportunities for work that are fairly compensated, stable, and distributed across the community and the business sector (i.e., not dependent on a single industry or commodity).

It is recognized that the community and its values will shift over time, as community composition changes along with health and other infrastructure changes. Shifts in community values will impact the extent to which restoration efforts are sustainable, perhaps in unexpected and difficult-to-predict directions. Improved economic conditions for the local people, either directly or indirectly through the mining operation, may reduce the impact on the local forests, as other sources of fuel and food can be obtained. An improved port, capable of handling large ocean vessels, will increase the accessibility of the region. Ecotourism is one economic endeavor in its infancy that is limited by the lack of local infrastructure, including the port.

Similarly, management practices for the restored areas will have to adapt over time, as all involved (QMM, local people, and authorities) learn how to best balance valuable ecological resources while improving the standard of living for local people. Adapting management approaches and tools, and perhaps even specific implementation objectives, may be necessary as the local area and surrounding region change and grow.

Case Study 3   193

C3.7 SIGNIFICANT OUTCOMES AND LESSONS

QMM has been able to develop and begin implementation of a plan that seeks to balance a number of conflicting interests and different values. The implementation outcomes over the next 5–10 years will provide significant lessons for other development projects in other parts of the world where the values associated with biodiversity may conflict with other values and needs. It is apparent that neither success in protecting biodiversity nor success in improving the standard of living of the local people will occur without planned economic development, and that success in either realm depends on concurrent success in the other.

Some of the more relevant lessons include the following:

- *Staff the environmental and social programs with full-time local employees.* It was recognized early on that the complexity of the issues required was too great to use outside consultants on an ongoing basis to solve the issues, and that both credibility and capacity building dictated a preference for in-country hires whenever possible.
- *Integrate environmental and social teams.* The teams were housed in the same building and encouraged to share ideas, plans, and results with each other regularly. This was seen as essential given the linkages between the key issues, such as the threat to the littoral forest and their biodiversity being largely due to destructive subsistence use of the forest. The integration has helped in ensuring that proposed solutions, projects, and activities are as multilateral as possible in addressing environmental and social needs.
- *Demonstrate to stakeholders that proposed restoration outcomes are feasible.* The results achieved from the research and fieldwork were able to show that plants of both biodiversity and commercial interest could be germinated and transplanted successfully, and that for traditional uses, the efficacy of the plants was undiminished.
- *Use outside experts to help address issues and solve problems.* Outside experts have proved invaluable to helping to address some of the issues associated with littoral forest protection and restoration, as well as the other aspects of the environmental restoration. Outside input is increasingly helpful in the social-cultural arena as well. Not only does this provide practical help, but also the inclusion of independent external individuals and organizations has helped in communicating the complexity of the issues and the need for integrated and creative solutions, and in building bridges and understanding with many who were critical of the project.
- *Complex problems cannot be fast-tracked.* Effective and sustainable solutions to complex problems require sufficient time to build trust and respect among relevant interested parties through deliberate consultation, to evaluate alternative approaches and solutions to multifaceted problems, and to demonstrate the feasibility of selected options.
- *Resource allocation must reflect potential impact on stakeholders.* The first priority for resource allocation was to understand the values and

address the issues of those living nearest the project. Then resources were allocated and deployed at the regional, national, and, finally, international levels.
- *Flexibility in planning is required for credible stakeholder engagement.* An effective and transparent stakeholder process meant that QMM had to be willing to change and adjust plans. For example, it was essential that the mining project fit into the vision for the overall regional development plan. The forum created to deliberate the regional development plan, in which QMM participated, provided an independent, public venue for issues and ideas to be debated. Through this process, it became clear that while there was agreement that a modern port suitable for large oceangoing vessels was required, the proposed port site favored by QMM would not be optimal for greater Fort Dauphin. QMM has modified its mining plans, and moved the future port location to one that meets the approval of other local and regional interests.
- *Continuous consultation is critical.* Consultation with relevant parties must occur regularly and often enough to ensure that people are informed of progress or changes to plans and that they have the opportunity to provide responses in time to have meaningful effect.

Case Study 4: 1991 Gulf War Oil Spill

Long-Term Impacts on Shoreline Habitats

Jacqueline Michel

CONTENTS

C4.1 "Scope" or Type of Problem: Impact Assessment 195
C4.2 Resource Management Goals and Objectives 196
C4.3 Stakeholder Identification 196
C4.4 Conceptual Model 197
C4.5 Values Identification (Assessment Endpoints) 197
C4.6 Data Requirements (Measurement Endpoints) 197
C4.7 Analysis Plan 198
C4.8 Characterization and Communication 198
C4.9 Lessons Learned 199
References 199

C4.1 "SCOPE" OR TYPE OF PROBLEM: IMPACT ASSESSMENT

During the 1991 Gulf War, there were intentional releases of an estimated 1 500 000 tonnes (450 000 000 gallons) of crude oil from 8 tankers, a major tank farm, and 2 offshore terminals (Tawfiq and Olsen 1993). It is the largest oil spill in history and 3 times as large as the next largest spill (the 1979 Ixtoc well blowout in the Gulf of Mexico). About 10% of the spilled oil was removed by on-water recovery, whereas there was very little shoreline cleanup. In Saudi Arabia, more than 800 kilometers of shoreline remained heavily oiled.

The oil spill was not cleaned up shortly after the spill because 1) there was a lack of a "responsible party" that could be required to pay for the cleanup; 2) there was no international mechanism to fund the immediate cleanup; 3) it was an act of ecoterrorism during a war that exceeded the ability of any one nation to pay for cleanup; 4) although a mechanism was eventually established to pay cleanup claims, cleanup was delayed for 10 years because the injured party was unwilling to implement the work and risk not being compensated; and 5) there was limited access to information about the extent of the impact and ways to correct it, despite a broad stakeholder demand that the spill be cleaned up.

Amazingly, no long-term impacts to subtidal habitats and communities were observed, including seagrass beds, coral patch and fringing reefs, unvegetated sandy and silty substrates, and rocky outcrops (Kenworthy et al. 1993; Richmond 1996). Little of the oil sank (Michel et al. 1993). Impacts to birds resulted from direct mortality (an estimated 30000 seabirds were killed) and severe declines in breeding success 1 to 2 years after the spill that were attributed to an acute shortage of food due to the oil impacting fish recruitment (Symens and Alsuhaibany 1996). During the spill, shorebird populations were reduced by up to 97%; however, it is not known whether the birds avoided the noxious oil or were driven away by a lack of food and found good feeding areas elsewhere, became oiled and died, or died from starvation (Evans et al. 1993).

The government of Saudi Arabia made a claim to the United Nations Compensation Commission (UNCC) for marine and coastal impact assessment studies to support future claims for remediation of oiled habitats and resources. Two of the studies dealt with intertidal habitats: 1) the Oiled Shoreline Survey, which was similar to a remedial investigation; and 2) the Treatment Technology Assessment, which was similar to a feasibility study for remediation of shoreline habitats.

C4.2   RESOURCE MANAGEMENT GOALS AND OBJECTIVES

The overall goal of the Oiled Shoreline Survey was to provide the justification for developing a remedial action plan. The objectives were to quantify the volume of oiled sediments remaining along the shoreline in Saudi Arabia and identify those habitats requiring remediation. Because this region also has chronic oil releases and previous spills, it was also necessary to confirm that the contaminated areas identified during the survey represent remnants of the 1991 Gulf War oil spills. Because natural recovery was a possible option, it was important to evaluate the ecological health of intertidal communities and identify the factors that were limiting ecological recovery.

C4.3   STAKEHOLDER IDENTIFICATION

The affected area is very sparsely populated and has restricted or very limited access. Saudi nationals traditionally recreate in the desert rather than along the shoreline, although there are beach resorts near population centers. There are a few fishing villages populated by foreigners. The primary stakeholder responsibility was assumed by the central government and without any public participation or transparency. On a regional scale, the northwestern Arabian Gulf shoreline is an important migratory stopover for shorebirds. As discussed below, these intertidal areas have been greatly impacted, significantly reducing their function as feeding grounds for migrating shorebirds and possibly affecting populations. Regional stakeholders would include those who value shorebirds in both summer and winter habitats. The ecological linkages to other ecological services and resources and their users in the Arabian Gulf (e.g., nutrient cycling, fish, shrimp, and fishers) have not been specifically identified but are assumed to be important.

Global stakeholders, such as environmental groups, were not involved because there was limited knowledge about the extent and duration of the impact.

C4.4 CONCEPTUAL MODEL

The project was based on the assumptions that the spilled oil in the intertidal zone had not been cleaned by natural processes, the oil continued to impact intertidal ecosystem health, and remediation was necessary to speed recovery to baseline conditions.

C4.5 VALUES IDENTIFICATION (ASSESSMENT ENDPOINTS)

The assessment endpoints were a combination of the accepted "oil spill cleanup" mentality that oil spills are emergency events that require removal of all "visible" oil (which was the endpoint used in the remediation claim) and a "Superfund site" mentality that persistent, highly contaminated habitats require remediation using risk-based criteria that would reduce impacts to ecosystems and human health (which will likely be the endpoints used for actual remediation). The issue of causing more harm during cleanup activities in sensitive salt marshes and mudflats must still be resolved.

C4.6 DATA REQUIREMENTS (MEASUREMENT ENDPOINTS)

Data were needed to calculate the location and volume of contaminated sediments by degree of contamination and by habitat. So that there would be little basis for contesting the study results, a very dense sampling plan was developed. Transects were established at 250-meter (m) intervals along more than 800 kilometers (km) of shoreline. Each transect started at the landward edge of the oil and ended at the seaward edge. Along each transect, trenches were dug, described, and sampled at intervals of 5 to 80 m. Standard visual descriptors on the degree of oiling were developed and validated with intensive sediment sampling.

The field teams completed 3107 transects, including 85 in unoiled "comparison" sites; dug, described, and photographed 19 515 trenches; and collected 26 158 samples for total petroleum hydrocarbon analysis, 2660 samples for detailed chemical characterization and fingerprinting, and 134 bivalve tissue samples. The ecological condition of the shoreline between transects was assessed using a Rapid Environmental Assessment protocol to measure species richness. The intertidal habitats were mapped using both the field data and remote sensing imagery.

The study results were needed to support the remediation claim, leaving only 5.5 months to complete the fieldwork. The Data Quality Objectives required a high level of data quality control and quality assurance (QA/QC). All field data were recorded using standardized data entry methods with pull-down menus on field computers linked to differential GPS receivers to record location and bar-code scanners for recording samples. Each night, all field data were downloaded, run through a series of QA/QC checks, and prepared for delivery to the client by the next morning. All samples were delivered under chain of custody to the laboratory daily. The analytical program was fast-tracked and managed under very tight QA/QC protocols.

C4.7 ANALYSIS PLAN

A phased data analysis approach for such a large, multicomponent data set was needed to meet the different decision-making requirements and deadlines. The first phase focused on calculating the oiled sediment by oiling category and habitat type using different estimation methods, to support the immediate deadline for the remediation claim. The next phase will be development of cleanup criteria so that site-specific remediation strategies can be determined.

C4.8 CHARACTERIZATION AND COMMUNICATION

Twelve years after the spill, 8 million cubic m of oiled sediments remained in the intertidal zone in Saudi Arabia. Almost 45% of the oiled sediments were on sheltered muddy tidal flats, and more than 23% were on salt marsh habitat. About 26% of the oiled sediments occurred on sandy tidal flats, and 11% on sand beaches. Thus, there had been limited removal by natural processes in the 12 years since the spill, and only in the areas of highest exposure to wave action and the lowest initial oil loading.

In the first phase of analysis, ecological health was measured using a species richness threshold based on the comparison sites. Nonrecovering habitats are those that have a species richness that is lower than the threshold. In these habitats, a shift (or disturbance) in community structure and species assemblage and composition was also observed. The sheltered habitats are the most impacted, with 80% of the salt marsh transects and 71% of the muddy tidal flat transects classified as nonrecovering. Recovering or disturbed habitats are defined as those that have a species richness that is equal to or greater than the threshold for that habitat. Reaching the species threshold is the first step of recovery. However, the species assemblage in most of the recovering transects is different from that of the comparison transects, indicating a disturbed or damaged community structure.

How to restore such large areas of sensitive habitat is a very complex issue. It requires an understanding of the many factors that affect the ecological recovery of the intertidal habitats that were oiled as a result of the 1991 oil spill. These factors include the following:

1. *The chemical toxicity of the oil residues.* Although the residual oil was characterized as moderately to extremely weathered, the levels of polynuclear aromatic hydrocarbons (PAHs) in 60% of the samples of sediments with visible oil exceeded the Effects Range — Low of Long et al. (1995) and thus are likely to have impacts to sensitive species. In 29% of the samples, the PAH levels exceeded Effects Range — Medium and thus are likely to affect a broad range of species. The sediments with the most potential toxicity are those that are the least weathered, which are located in the muddy sediments of the salt marshes and sheltered tidal flats.
2. *The physical toxicity of heavy and hardened oil residues.* Heavy accumulations of weathered oil residues on the surface can slow the recovery of intertidal communities even when the oil is no longer chemically toxic. Thick and hardened residues modify, prevent, or slow many ecological

processes, including changing the moisture content and flushing of underlying sediments, nutrient cycling, oxygen exchange, seed germination, settlement of larvae, feeding of intertidal grazers, and burrowing by crabs.
3. *Other physical barriers that affect seed germination of plants, settlement of larvae, and burrowing.* For example, algal mats are an important intertidal community along arid shorelines. However, in most oiled areas, the algal mat has expanded significantly. This expansion of thick algal mats likely plays a significant role in the delay of habitat recovery.
4. *Limited sources for recruitment of biota.* When large, contiguous stretches of habitat are so heavily impacted, recovery may be delayed because of limited sources of seeds and larvae for recruitment. This factor is particularly important for species that do not have pelagic eggs or larvae that can drift in from other areas.
5. *Hydrological functioning of tidal channels.* The tidal channels are not functioning as they would without the continuing effects of the oil and the presence of thick algal mats. Tidal channels are extremely important to the functioning of tidal flats and salt marshes, and they are a key to habitat recovery.

The role of these multiple factors limiting ecological recovery has to be evaluated, so that appropriate remediation strategies can be identified.

C4.9 LESSONS LEARNED

As this book goes to publication, the UNCC announcement has not been made public. Until then, it is not possible to describe accurately the lessons learned.

REFERENCES

Evans MI, Symens P, Pilcher CWT. 1993. Short-term damage to coastal bird populations in Saudi Arabia and Kuwait following the 1991 Gulf War marine pollution. Marine Pollut Bull 27:157–161.
Kenworthy WJ, Durako MJ, Fatemy SMR, Valavi H, Thayer GW. 1993. Ecology of seagrasses in northeastern Saudi Arabia one year after the Gulf War oil spill. Marine Pollut Bull 27:213–222.
Long ER, Macdonald DD, Smith SL, Calder FD. 1995. Incidence of adverse biological effects with ranges of chemical concentrations in marine and estuarine sediments. Environ Mgmt 19:81–97.
Michel J, Hayes MO, Keenan RS, Sauer TC, Jensen JR, Narumalani S. 1993. Contamination of nearshore subtidal sediments of Saudi Arabia from the Gulf War oil spill. Marine Pollut Bull 27:109–116.
Richmond MD. 1996. Status of subtidal biotopes of the Jubail Mainre Wildlife Sanctuary with special reference to soft-substrata communities. In: Krupp F, Abuzinada AH, Nader IA, editors, A marine wildlife sanctuary for the Arabian Gulf: environmental research and conservation following the 1991 Gulf War oil spill. Frankfurt a.M., Germany: National Commission for Wildlife Conservation and Development, Riyadh, Kingdom of Saudi Arabia and Senchenberg Research Institute, p 159–176.

Symens P, Alsuhaibany AH. 1996. Status of the breeding population of terns (Sternidae) along the eastern coast of Saudi Arabia following the 1991 Gulf War. In: Krupp F, Abuzinada AH, Nader IA, editors, A marine wildlife sanctuary for the Arabian Gulf: environmental research and conservation following the 1991 Gulf War oil spill. Frankfurt a.M., Germany: National Commission for Wildlife Conservation and Development, Riyadh, Kingdom of Saudi Arabia and Senchenberg Research Institute, p 404–420.

Tawfiq NI, Olsen DA. 1993. Saudi Arabia's response to the 1991 Gulf oil spill. Marine Pollut Bull 27:333–345.

Case Study 5: Example of Valuing the Ecological Benefits from the Clean Air Act and 1990 Amendments*

Brian T. Heninger
USEPA, National Center for Environmental Economics

CONTENTS

C5.1 Background 201
C5.2 Process and Results..202
C5.3 Conclusions 204
C5.4 Research Needs and Recommendations 204
References..205

C5.1 BACKGROUND

Over the years, the US Environmental Protection Agency (USEPA) has come under increasing pressure to better characterize both the benefits and the costs of the regulations it enacts and enforces. This is particularly true for regulations that have a significant impact on society, such as the Clean Air Act (CAA) and its amendments of 1990 (CAAA). Describing and quantifying all of the benefits and costs of this massive piece of regulation comprise a monumental task. However, the amendments of 1990 require the USEPA to periodically assess the effect of the CAA on the "public health, economy, and environment of the United States," and to report the results of its assessments to the US Congress.

In 1997, the USEPA completed its first Report to Congress, which presented a retrospective assessment of the benefits and costs of the CAA between 1970 and 1990 (USEPA 1997). The report showed that the benefits of the CAA over this 20-year period were approximately $22 trillion and the direct costs were $523 billion,

---
* The views expressed are those of the author and do not necessarily express those of the USEPA.

implying a benefit–cost ratio of 42:1. The majority (81%) of the assessed benefits were due to decreased incidences of premature mortality from particulate matter and lead. The remaining benefits of the CAA were due to reduced health effects, increased labor productivity from fewer restricted activity days, higher IQ in children, and improvements in visibility and agriculture. The benefits occurring to ecological systems were not quantified or monetized for this report.

Approximately 2 years later, the EPA released the results of a prospective Report to Congress, which was an assessment of the benefits and costs of the CAAA from 1990 to 2010, and found the total benefits to be $1.2 trillion and the costs to be $210 billion, implying a benefit–cost ratio of 6:1 (USEPA 1999). However, as with the previous Report to Congress, the majority of the assessed benefits (almost 90%) were due to reductions in premature mortality from particulate matter, with another 7% of benefits due to reductions in morbidity. The remaining quantified benefits were distributed between improvements in labor productivity, improved visibility at recreational sites, increased crop yields, and ecological improvements.

The ecological benefits estimated amounted to less than 1% of the total quantified benefits of the CAAA. However, this figure is not very meaningful. This is simply the sum value of the change in identifiable ecological service flows that result from the CAAA being in place that can be quantified and valued in dollar terms by assessors. This is a small subset of all the ecological benefits produced by the CAAA. Ecological service flows are those goods and services that ecosystems produce that are valued by society. More often than not, these are things that are not bought and sold in a market, such as groundwater filtration, biodiversity, and recreational activities. The greater heterogeneity in ecosystems services makes it relatively more difficult to produce estimates of the benefits from their protection than for the protection of human health. However, analysts do often have a need to put a dollar value on ecosystem services, just as they value health effects. The reason that dollars are typically the metric of choice for valuation is to facilitate the comparison with costs. This is, of course, the case in a benefit–cost analysis.

C5.2 PROCESS AND RESULTS

In order to value an ecological benefit attributable to the CAAA in dollars, several conditions must be met. The ecological endpoint must be an identifiable service flow from and/or within an ecosystem, a defensible link must exist between changes in air pollution emissions and the quality or quantity of the ecological service flow, quantitative models must be available in order to measure this relationship, and economic models must exist in order to monetize the endpoint. So, while there are a large number of known impacts of air pollution on ecological systems, only a subset of these can be assessed quantitatively given existing data and methods. Furthermore, only a smaller subset can then be valued in dollar terms, as it is only when changes in ecological systems can be quantified that economic modeling tools can then be used to estimate the value of those changes to society.

Case Study 5: Example of Valuing the Ecological Benefits

Ecological benefit estimation in the USEPA's second Report to Congress (1999) on the benefits and costs of the CAAA was also hampered by the fact that the CAAA is such a far-reaching regulation that an ecological assessment must be on a national scale in order to be complete. Given the existing difficulties of performing an ecological assessment of an air pollutant stressor at a particular geographic site, one can envision the hardships involved with undertaking a national assessment of multiple air pollutant stressors.

Of the many ecological improvements due to the CAAA, only 2 were quantified and valued in dollars. The first was the increased commercial timber growth from reduced ozone levels. This was quantitatively assessed on a national scale. The existence of quantitative methods and models for estimating the change in net primary productivity of the various tree species made this possible. Furthermore, valuation of the change in this service flow was relatively straightforward, as the value of increased timber growth can be estimated by observing market prices. Specifically, the Timber Assessment Market Model, an economic model of the forest sector, was used to calculate the value of the estimated change in increased forest growth nationally.

The only other ecological improvement due to the CAAA that was included as a monetized benefit in this final Report to Congress was decreased acidification to freshwater lakes and streams in the Adirondack region. However, decreased acidification is not an ecological service flow that is directly valued by society. An ecological endpoint that is valued directly is the increased willingness to pay anglers to fish in this region due to the improved quality of the lakes and streams. In this case, quantitative methods existed to estimate the ecological impacts avoided from decreased acid deposition on a national scale, but economic models capable of valuing society's increased willing to pay to achieve this (for the purpose of improved fishing) were only available for specific regions.

The Report to Congress also contained a quantitative assessment of decreased nitrogen deposition into east coast estuaries. Additionally, carbon sequestration due to increased commercial timber growth was assessed quantitatively. However, valuing these physical changes in dollars proved too difficult. For example, decreased nitrogen deposition improved the health of submerged aquatic vegetation in estuaries, which bolstered fish populations, thus improving recreational and commercial fishing. However, these causal links could not be quantitatively established for the change in question. Establishing a defensible quantitative link between decreased emissions and ecological effects proved too difficult in many cases, and when this problem was overcome, the economic models necessary to link an ecological service flow to a change in ecosystem function were not available. For example, estimating the accumulation of toxics in freshwater fisheries was limited by the lack of toxics deposition and exposure data. Aesthetic degradation of forests associated with ozone and airborne toxics exposure was observable, but the link between aesthetic degradation and specific air quality scenarios was uncertain, and studies of society's willingness to pay to avoid foliar damage were only available for specific geographic areas. A long list of beneficial ecological effects occurring as a result of the CAAA was produced and characterized qualitatively as part of the report.

C5.3 CONCLUSIONS

The benefit–cost assessment of the CAAA illustrates many of the difficulties in valuing the ecological benefits of a major national environmental regulation. The linkages of cause and effect between air pollution and an ecosystem's structure and function are difficult to quantify. This is particularly true for subtle and/or long-term deterioration in ecosystem integrity. Additionally, degradation of ecosystem integrity most often does not cause immediate, measurable declines in ecosystem service flows that are normally valued by society. For example, the decline of a little known plant or insect species population in a particular ecosystem rarely has short-term economic effects on markets or societal welfare. The absence of clear methods for accomplishing the modeling of ecosystem effects and the consequent economic implications does not imply that ecological impacts with no immediate consequences are of no economic concern. This is a major conceptual barrier in the quantification of economic benefits associated with nutrient cycling, water filtration, biological diversity, and provision of habitat.

Changes in ecosystem structure and function are sometimes unpredictable in nature and scale. Ecosystems affected by air pollution may respond in a discontinuous manner around critical thresholds that are boundaries between locally stable equilibria. Complexity in ecosystems prevents analysts from using linear methods to "add up" the discrete ecological effects of pollution. Understanding the complex cause-and-effect relationships between air pollution and deterioration of ecosystem structure and function is fundamental when performing this type of analysis. The isolation of ecological service flows may often imply an oversimplified cause-and-effect relationship between pollution and the provision of the service flow, when more often the service flow is affected by complex nonlinear relationships that govern ecosystem structure and function. The result is that ecosystem impacts may not be adequately assessed by analyses that focus on specific service flows (USEPA 1999).

C5.4 RESEARCH NEEDS AND RECOMMENDATIONS

There is a need for more assessments with a greater emphasis on ecosystem integrity and use of ecosystem-level models, in place of oversimplified dose–response functions of ecological service flows. There is also a need for assessments with a broader geographic coverage and more studies that allow for transferability of results across localities. Additionally, many studies could benefit from a more sophisticated treatment of uncertainty in both the assessment of ecosystem impacts and economic valuation. Furthermore, increased synergy between ecological assessments and economic valuation would increase the likelihood that all relevant ecological effects are properly recognized, quantified, and used as inputs in economic valuation models. This increased collaboration between disiplines would increase the likelihood that economic valuation models are developed for the ecological attributes ecologists can assess and, conversely, that they assess ecological endpoints for which economic valuation models exist.

Finally, in an assessment such as the benefit and cost analysis of the CAAA (Report to Congress; USEPA 1999), the incomplete quantification and monetization of ecological benefits lead to a downward bias in the total monetized benefits of the regulation. This fact must be indicated to the reader up front to avoid any misinterpretation or misuse of the benefit–cost numbers from this study. One should keep in mind that benefit–cost analysis is a tool to aid decision makers, not the sole tool used to reach a decision. In this case, numerous benefits to various ecosystems' structure and functions were identified and characterized qualitatively; and this information is part of the analysis, even though the results do not carry though to the final benefit–cost figures. Additionally, as a result of this study and other motivations, more research and resources at the USEPA are going into the identification, quantification, and monetization of ecosystem functions and services.

REFERENCES

[USEPA] US Environmental Protection Agency. 1997. The benefits and costs of the Clean Air Act, 1970 to 1990: EPA report to Congress. EPA-410-R-97-002. Washington (DC): USEPA, Office of Air and Radiation, Office of Policy, Planning and Evaluation.
[USEPA] US Environmental Protection Agency. 1999. The benefits and costs of the Clean Air Act, 1990 to 2010: EPA report to Congress. EPA-410-R-99-001. Washington (DC): USEPA, Office of Air and Radiation, Office of Policy, Planning and Evaluation.

Case Study 6: The CALFED Bay–Delta Program
A Case Study in Intertwined Ecological and Environmental Resource Issues

Samuel N. Luoma

In California, whiskey is for drinkin' and water is for fightin' over.

Paraphrase of Mark Twain, circa 1870

CONTENTS

C6.1 Scope of the Problem 207
C6.2 Resource Management Goals and Objectives 209
C6.3 Stakeholders 210
C6.4 Conceptual Model 211
C6.5 Values Identification 212
C6.6 Data Requirements 213
C6.7 New Knowledge: Learning and Managing Adaptively 214
References 215

C6.1 SCOPE OF THE PROBLEM

The California Bay Delta region extends from headwaters high in the Sierra Nevada Mountains to where the Sacramento and San Joaquin Rivers flow into San Francisco Bay and on to where the bay enters the Pacific Ocean. The modern delta is composed of a complex web of waterways, dikes, and channels with some small freshwater tidal lakes and wetlands. An estimated 700 miles of sloughs and waterways surround 57 manmade islands. The bay is an estuary with extensive mudflats and a few hundred acres of saltwater and freshwater wetlands and marshes (less than 10% of the original wetland areas remains undisturbed). The watershed covers approximately 40% of the area of California and carries 60% of the state's water supply.

Controversies over California's water are of legendary proportions (see Roman Polanski's film *Chinatown*). The entire state is semi-arid, with strongly seasonal precipitation (snow and rain in the winter and spring; no precipitation during the summer and fall) and a serious spatial discontinuity between the water supplied by nature and the demands from human activities (most water is in the north, and most demand is in central California and, especially, the south). Beginning in the 1930s, the system was engineered to harness the water supply and counter nature's temporal and spatial discontinuities. Water is captured in dams during the wet season (every major stream and river is dammed, except one). That water is then moved toward the delta, where it is pumped uphill and distributed during the dry season to agricultural interests in the southern Central Valley and the cities of southern California. Thirty-five percent of the Sacramento River can be diverted during the wet season, and 60% of the Sacramento River plus all of the San Joaquin River can be diverted during the dry season (Hayes 2002). Nichols et al. (1986) called this the largest water management system in the world. The volume of water exported for agriculture and urban interests increased progressively from the 1930s until the early 1990s.

Development and operation of the water management infrastructure stressed the aquatic environment in numerous ways. Access of Chinook salmon to half of their historic habitat was eliminated by dams. Rivers and streams were dewatered by diversions, and channelized to move water more effectively from north to south. Dewatering and channelizing raised temperatures during critical times of year in many tributaries, eliminated floods that are critical to resetting stream geomorphology, and eliminated in-stream habitat, wetlands, and floodplain habitat. All these stresses had implications for salmon and for native fish that depended upon spring floods, cold water, floodplain breeding habitat, and the nursery functions of wetlands. Fish were also removed directly from streams by thousands of diversions along the rivers in the delta; and massive fish mortalities were observable during some times of year at the large pumping facilities in the delta. Other pressures on aquatic ecosystems also grew during the development of the watershed, as a result of urbanization, agricultural pollution, industrial pollution, mining (both historic mineral mining and the modern extraction of gravels from streams), invasions by exotic species, and commercial fishing (Nichols et al. 1986).

After an 8-year drought that extended from 1984 to 1992, dramatic declines in the abundance of some fish species were evidenced as indicators of the serious environmental implications of human development (see, e.g., Figure C6.1; see also Nichols et al. 1986; Hollibaugh 1996). The general consensus among scientists studying the system was that the water diversions and the water supply infrastructure provided the single biggest source of ecological stress in the system, although quantitative evidence defining the relative importance of different stressors in different circumstances and for different species was lacking.

The passage of federal and state environmental laws (the Clean Water Act and the Endangered Species Act [ESA]) and increasing public interest in the environment had the effect, in the 1990s, of changing the trajectory of growth in the diversion of water to satisfy mostly agricultural and urban interests. In the early 1990s, winter-run Chinook salmon were declared endangered under the ESA, and other species were soon listed as well. Government agencies charged with enforcing the act were thereby empowered to take all means to protect these species, whatever the costs. Regulations were passed requiring first choice of some water to go to the agencies that managed migrating fish. They specified that enough water pass down

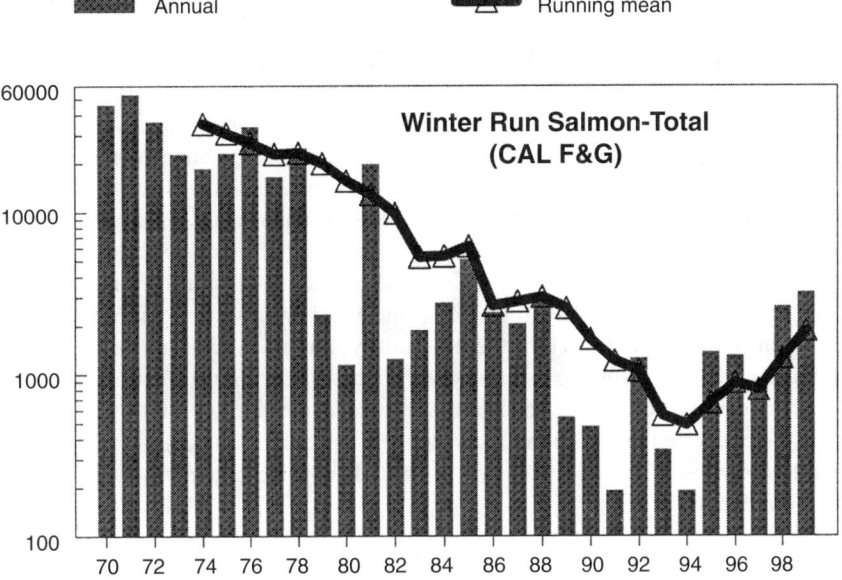

FIGURE C6.1 Population size (number of adult individuals) of winter-run Chinook salmon in the Sacramento River watershed. *Source:* Data from California Fish & Game (2004).

the Sacramento River to maintain a salinity standard in the estuary and dedicated a specific parcel of water to the agencies to manage in the name of protecting migrating fish (800 000 acre feet, compared to about 6 to 10 million acre feet contracted for diversion south). The agencies were allowed by the courts to shut down the large pumps in the delta whenever winter-run or other species listed under the ESA were migrating through the area. These activities caused a near crisis for California water management agencies and water users (Jacobs et al. 2002). At that point, stakeholders traditionally not on speaking terms saw the need to find common ground.

C6.2 RESOURCE MANAGEMENT GOALS AND OBJECTIVES

All stakeholders eventually saw that the deadlock between water user interests and environmental protection interests required a solution. The goals that all eventually agreed upon were formally stated as follows:

Ecosystem quality. Improve and increase aquatic and terrestrial habitats, and improve ecological functions in the Bay–Delta to support sustainable populations of diverse and valuable plant and animal species.

Water supply. Reduce the mismatch between Bay–Delta water supplies and current and projected beneficial uses dependent on the Bay–Delta system.

Water quality. Provide good water quality for all beneficial uses.

Levee system integrity. Reduce the risk to land use and associated economic activities, water supply, infrastructure, and the ecosystem from catastrophic failure of delta levees.

The Programmatic Record of Decision (ROD; CALFED Bay–Delta Program 2000) that formalized the solutions was signed in 2000 and specified approximately 50 actions to be evaluated over a 7-year period and implemented over a 30-year period. A unique aspect of the plan is that it formally requires a lead scientist, an independent scientific advisory board, the use of world-class science, and the use of adaptive management. The specific actions included ecosystem restoration, additions to the water storage system, and large engineering changes in the delta that would improve the efficiency of moving more high-quality Sacramento River water to the south (sometimes in exchange for low-quality San Joaquin River water). In sum, these actions amounted to trade-offs designed to both improve management of the reliability of an ecosystem service (water) and improve the functioning of the ecosystem (Hayes 2002). While there certainly are (potentially at least) inherent conflicts in satisfying both goals, this type of trade-off dilemma is increasingly common in modern societies (van Eeten and Roe 2002). One question for this publication is whether valuation of environmental resources can aid in determining trade-offs or degrees of trade-off that might help decision makers better reach solutions appropriate to both goals.

C6.3 STAKEHOLDERS

The CALFED Bay–Delta Program was formed in 1994 under the leadership of then US Secretary of the Interior Bruce Babbitt and then California Governor Pete Wilson. The CALFED assignment was challenging: to develop a collaborative state–federal management plan for the system and involve multiple stakeholders whose interests frequently were in direct conflict (primarily municipal, agricultural, and environmental interests). Although many other resource management issues involve multiple stakeholders and conflict is integral to their discussion, the CALFED experience is unique due to its shared state and federal roles, the magnitude and significance of stakeholder participation, and the complexity of the scientific issues involved. The resulting CALFED Bay–Delta Plan (CALFED Plan) was a cooperative agreement crafted by stakeholders and more than 20 state and federal agencies with management and regulatory responsibilities for the Bay–Delta ecosystem. This long-term, comprehensive plan formalized in an ROD and associated Memoranda of Understanding signed by all state and federal agencies (CALFED Bay–Delta Program 2000).

The intensity of the involvement of multiple interests in the program on a day-to-day basis is evidenced by the reporting requirements of the program staff. They report to a state–federal agency management team (senior managers from the 22 agencies) that meets weekly, a board of state and federal leaders including the governor of California and the secretary of the interior that meet quarterly under the auspices of the legislated California Bay–Delta Authority, and a Bay–Delta Public Advisory Committee, run by stakeholders, that meets quarterly. Stakeholders are participants in many of the committees that make day-to-day decisions (e.g., a weekly meeting that tracks migratory fish movement through the system and recommends management actions to leaders on a day-to-day basis). Transparency of activities with public

Case Study 6: The CALFED Bay–Delta Program   211

reviews and advice from experts (usually well-credentialed academics) are goals for every major aspect of the CALFED program, and for at least some aspects, these have become the way of doing business (Jacobs et al. 2002).

C6.4 CONCEPTUAL MODEL

Development of common conceptual models is a critical ingredient in all individual CALFED actions. For example, every restoration proposal submitted for funding (most such awards are competitive) must include a conceptual model. At the scale of the major trade-offs, however, an explicit model has not been published. Nevertheless, implicit trade-offs are an important aspect of almost every element of the plan and were surely considered in developing the compromise. However, it is not clear that the formal use of economic valuation tools was common when the accord was being drawn up.

Conceptualization of a few of the proposed actions that involve trade-offs, and the net benefits (+) or costs (–), as superficially perceived with respect to 3 major goals (Table C6.1). The ultimate accord presumably resulted when the different interests all found the net outcome of all trade-offs acceptable. Presumably, the balance in any one set of trade-offs was less important than the balance among all. It is also noteworthy that more complete analyses and advancement in technical understanding have since showed that net benefits or costs were sometimes counterintuitive (not as perceived) or could be changed by new advances in knowledge. As a result, CALFED is confronted with the necessity of reconsidering deadlines, some details of some actions are changing through time, and it appears likely that some actions will not be implemented at all.

TABLE C6.1
Trade-offs in the CALFED Bay–Delta Program: Examples of actions and perceived net cost or benefit of each

	Restore species	Better quality drinking water	Reliable supplies
Improve fish access to habitat (e.g., remove dams)	+	0	0
Restore gravels in rivers and streams	+	0	0
Restore natural flow regimes in rivers	+	– or 0	–
Increase storage in reservoirs (build and raise dams)	– to +	+	+
Reengineer water movement through delta	–	+	+
Change diversion strategies (daytime only versus 24 hours per day; only some tidal phases)	+	+	+ to –
Divert more Sacramento River water	–	+	+
Increase volume of mandated "environmental water"	+	–	–
Bank water in environmental account	+	+	+

It is interesting to think about where and how formal valuations might best contribute in an environment like this. How would complexities in the system, new knowledge, or the importance of considering (and continuously reconsidering) the balance of all valuations affect the feasibility and usefulness of such valuations? Another interesting question is whether formal environmental valuations would have made any difference in the ultimate deal. In the following sub-section, one overarching trade-off (arguably the most important one) will be considered.

C6.5 VALUES IDENTIFICATION

A major part of the accord was to trade investment in the full suite of environmental improvement measures for assurances that water supplies would not be curtailed when listed species were threatened with "take" at the pumps in the delta. Implicit identification of values related to this trade-off occurred at more than one point in time in the development and implementation of the CALFED Plan. The first was when government environmental managers stopped water exports to protect the endangered winter run. The availability of data from decades of monitoring greatly reduced uncertainties about the potential for extinction of winter-run Chinook salmon (Figure C6.1). It was quite certain that hundreds (~200 to 300) of adult fish returned to spawn in 1991 and 1994, and that tens of thousands of fish (> 50 000 adults) were migrating in the 1970s. There was also little uncertainty that winter-run salmon were even more abundant before human perturbations began in 1850 (Figure C6.1).

Attempts to place a monetary value on a major charismatic species on the edge of extinction would certainly have been controversial and, arguably, not very useful in the early 1990s. The solutions chosen by the agencies were controversial but viewed by most as defensible: disrupt water supplies to eliminate mortality at the pumps, regulate some river inputs to the estuary, and reduce immediately manageable stresses like harvest.

In 1992, the drought ended and ocean conditions entered a cycle more favorable to Chinook salmon. It is also likely that the protective actions began to take effect. Winter-run Chinook populations began to rebound in 1994 (Figure C6.1). In 2003, approximately 20 000 adults returned to spawn.

The ROD (final agreement) was, as mentioned above, signed in 2000. It included a compromise to continue the protective actions for winter-run salmon (and other endangered species). The essence of that compromise was a promise of $150 million per year for 7 years for "ecosystem restoration" and a designation of ~$50 million per year to purchase additional environmental water that was under the control of the agencies managing biological resources. This water was used to replace water lost to users when the exports were curtailed to protect migrating species. Thus, in exchange for restoration and environmental water, agriculture and urban interests were assured of a reliable water supply. Although the compromise continued to hold through 2006, questions relevant to valuation continue to be raised.

C6.6 DATA REQUIREMENTS

A number of questions about data are raised by the implicit valuation in the agreement, many of which are commonly seen as considerations or limitations in the valuation literature.

Currency. Did the ecosystem advocates deal in the right currency? Both water (in the form of regulations) and money were received to help with ecosystem restoration. Later analyses suggest that other possibilities for currencies might have been even more beneficial. For example, scientific review suggested that it might be more effective to use environmental water to enhance flows in tributary streams rather than just in the delta. Valuation of the water itself in monetary terms probably would not have influenced outcomes, however.

Relative values traded off. Are the amounts of water and money per year correct to balance potential adverse effects and actually lead to the restoration of species? Are they too much (as some water use advocates claim) or too little? There is no evidence that the environmental agencies valued the environmental resource in a formal way before they reached their position. But there might have been an implicit valuation by water users based upon the costs of export disruptions and the benefits from a higher likelihood of meeting anticipated needs in the future. A large part of the ecosystem investment was also public money in the form of appropriated funds and, especially, funds (~$4 billion) from bond issues passed in initiatives for ecosystem restoration and clean water during 4 elections beginning in 1996. One can speculate that the availability of public money made it easier for water users and managers to compromise.

A second type of valuation question is raised by how the ecosystem agencies have chosen to spend the monetary investment in ecosystem restoration. Ecosystem restoration remains an art for which the future benefits are not always predictable. Where multiple stressors are important (which is usually the case), prioritization of investments is a concern. One school of restoration managers argues that restoration of full ecosystem function is the desirable focus. Other managers prioritize investing in recovery of species under the greatest threat (investments emphasize the needs for those specific species).

CALFED diversified its portfolio of investments. About $100 million per year was spent on ecosystem restoration and supporting scientific studies. This is essentially divided among several threats (environmental pollution, damaged habitats, reduced flows in rivers and streams, and isolated floodplains). Some of that restoration is targeted for the needs of threatened fish, but most benefit ecosystems as a whole. While consistent with the consensus among restoration scientists, this strategy proved vulnerable to political criticism for being too diffuse.

Progress from investments in both restoration and environmental water proved difficult to monitor. Populations of some fish (e.g., salmon) increased, but others did not, during the first decade of actions. But it was difficult to attribute successes and failures to any specific action or stress. A valuation of how the investment matched the value of the recovering resource would suffer from the problem described above (valuing a species on the edge of extinction) and suffers from the uncertainties in the ecological data. Nevertheless, even a rough estimate might be a useful way to assure a balanced or systematic evaluation of the program.

Two other stressors were chosen for massive investment. Approximately one-third of the $150 million per year restoration funds goes to help screen diversion points in the rivers and delta. Thousands of these points exist, ranging in size from a few cubic feet per second (cfs) to thousands of cfs. The agencies stipulated that all these points must be screened, so ~$25 million to $50 million was spent building such screens. Although it is clear that fewer fish end up in diversion canals when diversion points are screened, the agencies have chosen not to study crucial questions like what kind of fish are protected by such screens, how many are protected from loss by this kind of investment, and what the relationship is between the amount of water diverted and the loss of native or threatened species. The largest diversion points were screened by 2003. Important questions were then raised about continuing this investment. In this case, economic valuation seems especially well suited for comparing the cost of investment in screening small diversions to the value of the fish saved by screening (even if the value of the latter has an arbitrary aspect). But collecting better ecological data on the specific biological benefits would be an essential prerequisite.

Finally, $50 million per year is spent on banking environmental water, purchased with public funds, then releasing it for the water users when pumping in the delta is curtailed to prevent massive mortalities of threatened species. This is an experiment in managing adaptively that will be evolving as new data arise and new actions are taken. Superficial valuations were occasionally used in discussions about how this water should be deployed or whether the investment is worthwhile. A systematic valuation exercise might be especially valuable in this case.

C6.7 NEW KNOWLEDGE: LEARNING AND MANAGING ADAPTIVELY

Salmonids are not the only species in the Bay–Delta system listed under the Endangered Species Act. Other species include delta smelt, which is a small, weak-swimming indigenous animal found only in the bay watershed. Even though it has no known commercial value, its extirpation from this system would result in an irreversible, global extinction of the species. Delta smelt may prove to be a useful prey species in the salmon food web, but those interactions are not well known.

Because the delta smelt life cycle is poorly known, controversy also exists as to the magnitude of the threat posed by water diversions to this species, CALFED gave a high priority to support for delta smelt research, and the research community produced results of value to environmental managers. Recent discoveries show apparent direct benefits to this species from curtailing water diversions when large populations migrate toward the pumps. The size of the delta smelt population in any 1 year also appears dependent on the number of spring days below a threshold air temperature. Cooler years yield longer reproductive periods and thus more delta smelt.

Warm years (like those typical during El Niño events) yield smaller populations. It follows that warm years make the total smelt population vulnerable to catastrophic decline, especially if combined with atypical mortality events from human activities (e.g., mass deaths at the export facilities). Populations of delta smelt fluctuated wildly between 1997 and 2005. Very low numbers in 2003 to 2005 contributed to a general cynicism about the success of the CALFED Program as a whole by 2005. Thus, a species that is almost impossible to value in monetary terms probably had the greatest impact on the political and practical acceptability of the program from both political and practical viewpoints.

Advances in science between 1997 and 2003 showed that proving benefits of curtailing water exports was difficult for both salmon and delta smelt. Uncertainties in ecological data provided the most important bottleneck. Nevertheless, advances in knowledge have influenced management. Salmon managers seem much more amenable to considering the use of environmental water to ameliorate multiple stresses on the population. For example, one novel proposal was to continue to use curtailment of exports to manage delta smelt, but to worry less about curtailing exports when salmon are near the pumps, and use that volume of water upstream to improve flows in streams where flows are known to limit salmon survival. Another is to focus on screening salmon at the large diversions, but not to proceed with the extremely expensive proposal to use screens to protect delta smelt (a proposal of questionable feasibility). Curtailed pumping would remain the dominant strategy for the latter. These new management proposals raise interesting possibilities for (perhaps) novel uses of environmental valuation. For example, such tools seem useful for comparing the value of water used (or lost to use) upstream of the delta versus curtailing water use in the delta (the comparison involves only feasibly monetized resources).

California water managers have a history of sometimes changing their approaches in response to new knowledge, although science is rarely given credit for such changes. If they do that here, it would reflect the best of a dynamic system that builds new knowledge and manages adaptively as that knowledge proves itself useful. Time will tell whether these are traits of the California system.

Not only are ecosystems dynamic, but our knowledge of ecosystems is also very dynamic. This is partly because environmental science is a young field and we are only beginning to find ways to communicate that knowledge to managers. This dynamism should be a caution to economists to be sure they are using the latest and best information in their analysis and to not expect knowledge in ecology to remain static for very long. More important, new management options may offer many new opportunities for ecology and economics to link in comparing the economic costs and benefits of the options.

REFERENCES

CALFED Bay–Delta Program. 2000. Programmatic record of decision: August 2000. http://calwater.ca.gov/Archives/GeneralArchive/RecordOfDecision2000.shtml. Accessed May 4, 2007.
California Fish and Wildlife. 2004. Sacramento River winter-run Chinook Salmon: biennial report 2002–2003. http://Qfg.cagov/nafnb/pubs/2004/ChinookWR0203.pdf. Accessed August 16, 2007.

Hayes JT. 2002. Federal-state decision-making on water: applying lessons learned. Washington (DC): Environmental Law Institute. 154 p.

Hollibaugh JT, editor. 1996. San Francisco Bay: the ecosystem. San Francisco (CA): American Association for the Advancement of Science. 535 p.

Jacobs KL, Luoma SN, Taylor KA. 2002. CalFed: an experiment in science and decision-making. Environ 45(1):30–41.

Nichols FH, Cloern JE, Luoma SN, Peterson DH. 1986. The modification of an estuary. Science 231:567–673.

van Eeten MJG, Roe E. 2002. Ecology, engineering and management: reconciling ecosystem rehabilitation and service reliability. New York (NY): Oxford University Press. 268 p.

Index

A

Abiotic environmental processes, 104
Abstraction, by scientists *vs.* economists, 168
Abundance, natural cycles of, 123
Accountability, in management process, 132
Actors, properties of, 13
Adaptability, in management process, 132
Adaptive management, CALFED Bay-Delta Program, 214–215
Adaptive Methodology for Ecosystem Sustainability and Health (AMESH), 95
Aesthetic judgments, *vs.* preferences, 5
Aeta people, 48, 51
　ancestral lands claim by, 178
　maintenance of cultural identity, 175
Aggregation
　difficulties with compliance data, 144
　and heterogenicity capture, 74
All or none decisions, 19
Allocation of resources, 45
Alternative livelihood strategies, in Subic Bay redevelopment project, 178
Alternatives, comparison of, 94
Altruism, values based on, 10
Ambiguity, 79
　and uncertainty, 78
Analysis approaches, 86, 133
　Gulf War oil spill, 198
　PCB-contaminated rivers valuation, 182–184
　Subic Bay redevelopment, 177
Analytical scale, 101
Animal welfare, 55
Appalachian Regional Assessment, 139
Applegate Watershed Partnership, 144
ArcView GIS, 96
Aroclor, 179, 180
Asbestos, public health hazards, 81
Assessment endpoints, 146–147
　comprehensive and systematic approach to identifying, 146
　data collection and DQO process in, 156–157
　Gulf War oil spill, 197
　identifying, 149–150
　identifying relevant ecological properties for, 150–154
　PCB-contaminated rivers valuation, 182
　problem identification and goal setting, 147–148
　stakeholder identification and, 148–149
　using ecology and economics to identify, 157–159
Asset values, as derived values, 62
Attitude surveys, 43
Attribute-based methods, 64
Attributes, complexity of quantifying, 103
Audio recordings
　of focus group sessions, 41
　qualitative data from, 31

B

Bald eagle, resource management of, 20
Baseline, in HEA method, 68
Basins of attraction, 95
Bayesian analysis, 84, 94, 106, 107, 118
　and uncertainty, 78
Behavioral context, 13
Benefits. *See* Cost-benefit analyses
Bentham, Jeremy, 51
Benzene, public health hazards, 81
Beyond compliance, 21
Bioaccumulation models, 99, 114
Biochemical oxygen demand (BOD), 72
Biodiversity, 151
　compatibility with mining project, 189
　conflicts with local uses, 156
　valuation in Madagascar case study, 119–122, 187–194
　values associated with, 3
Biological resources, 176
Biomes, as type, 102
Biosphere gas cycling, 69
Biospheric altruism, 10
Biotic-abiotic relationships, 105
Biotic responses, nonlinear, 104
Birds, integration of ecology and economics, 111
Blue crab controversy, 18–19
Boundaries, 100
　of ecological uncertainty, 125–126
　establishment for Subic Bay redevelopment, 176
Bovine spongiform encephalopathy (BSE), 80
Bright line, in Kantian ethics, 12, 19
Bruins, Randall J.F., xv
Bureau of Land Management, 141
Business impacts valuation problem, 21

217

C

CALFED Bay–Delta Program, 18, 19, 67, 157, 166
  case study, 207
  conceptual model, 211–212
  data requirements, 213–214
  incorporation of new knowledge for, 214–215
  resource management goals and objectives, 209–210
  scope of problem, 207–209
  stakeholders in, 210–211
  system complexity in, 122–124
  trade-offs in, 211
  values identification, 212
Canadian International Development Agency, 119, 188
Capital assets, measuring as stock, 62
Carbon sequestration, value of, 118
Carcinogens, FDA regulations for management of, 115
Cardinal measures, 81–82
Case studies, 7, 40
  CALFED Bay-Delta Program, 122–124, 207–215
  Clean Air Act ecological benefits valuation, 201–205
  Gulf War oil spill, 195–199
  Madagascar extractive development, 187–194
  PCB contamination, 124, 179–185
  Rio Tinto mining in Madagascar, 119–122
  Subic Bay redevelopment, 173–178
Causal loops, 95
Chesapeake Bay, blue crab case history, 18
Chinatown, and CALFED Bay-Delta Program, 207
Chinook salmon, 19
  CALFED case study, 122–124
  population size in Sacramento River watershed, 209
  progressive decline of, 123
  protection by CALFED Bay-Delta Program, 18
  threats to habitat, 208
Chlorine contamination, 179–180
Choice experiments, 64
Clean Air Act, 7
  problems in quantifying benefits of, 53, 166
  research needs and recommendations, 204–205
  valuation process and results, 202–203
  valuing ecological benefits form, 201–205
Clean water, methods for inferring value of, 66
Clean Water Act, 17
  and Chinook salmon crisis, 209
Clearly delimited questions, 29
Closed appraisal approaches, 61–62
Cognitive interviews, 32
Collective consciousness, 14
Collective valuation, 14
Command and control measures, 47
Commercial development, hypothetical case study, 116
Commercial forest crops, in Madagascar case study, 121
Common level of analysis, 101
  seeking, 100–101
Communities
  definition in space and time, 29
  identifying, 40
  identifying appropriate, 114
  in Madagascar case study, 187–194
  spatial *vs.* ecological definitions, 15
  as types, 102
Community, examples of defining, 101
Community-based conservation, 22
Community needs, 28
  in Madagascar case study, 119
  valuation in Madagascar case study, 119–122
Community relationships, 28
Community values, 14, 15
Compensation, for individuals, 52
Compensation test, 49
  as basis for economic valuation, 50
Complex systems, 107
  Everglades National Park, 170
  variability and uncertainty in, 103–106
Complexity, 75, 76, 165
  ability to incorporate, 77
  appreciating and understanding, 125–126
  in CALFED case study, 122–124
  confronting, 100–107
  in ecological systems, 97–100
  impossibility of fast-tracking, 193
  inherence in ecological systems, 98
  in Madagascar mining development, 119
  obfuscation by political pressure, 99
Comprehensive Environmental Response, Compensation, and Liability Act (CERCLA), 124, 132
  relationship to NRDA, 185
  remediation in PCB-contaminated rivers valuation, 182
Conceptual models
  benefits of clearly defined, 125
  CALFED Bay-Delta Program, 211–212
  for decision analysis, 93
  for ecological decision-making, 5
  Gulf War oil spill, 197
  and level of analysis, 102–103
  Madagascar case study, 189–190

Index

PCB-contaminated rivers valuation, 181–182
   scale and ecological type in, 102–103
   Subic Bay redevelopment, 174–175
Consequential surveys, 63
Conservation International, 119, 188
Constraints, and tractability, 103
Construction projects, 18
Constructive technology assessment, 72
Consumers, 152, 154
Contaminant concentrations
   integration of ecology and economics, 113
   in PCB case study, 124
Context
   responsiveness to, 75, 76, 82–83
   and values of interest, 14–15
Contingent valuation, 63–64
Continuous consultation, in Madagascar case study, 194
Contractarianism, 51–52, 55
Contradictory certainties, 80
Control, 154
Cost-benefit analyses, 44, 46, 48, 61, 67, 137
   for Clean Air Act benefits valuation, 204
   in economist–ecologist joint valuation example, 116–117
   in environmental decision-making, 2
   goal for public decision makers, 138
   implementation of compensation test by, 49
   practical problems in measurement, 52–54
   of uncertain future losses from extinction, 117–118
   United States debates over, 2
Creel surveys, 113
Cross-disciplinary integration, understanding of complexity as prerequisite for, 170
Current period, in valuation studies, 116
Current preferences, fallacy of basing future value on, 117
Curse of specificity, 15
Curtailments, ecological benefits of, 122
Cycles of abundance, 123
Cycling processes, 151

D

Data collection, 26–29
   CALFED Bay-Delta Program, 213–214
   Gulf War oil spill, 197
   methods for eliciting qualitative data, 32
   methods for eliciting quantitative data, 33
   processes, 30–31, 40
   qualitative and quantitative data, 31–32
   value of multiple methods, 33
Data Quality Objectives (DQO) process, 156, 157

Data requirements
   PCB-contaminated rivers valuation, 182
   Subic Bay redevelopment, 176–177
Decision analysis frameworks, 77, 93, 126, 171
   adaptive methodology for ecosystem sustainability and health, 95
   applications of, 94–96
   Bayesian analysis, 94
   conceptual approaches, 93
   Ecosystem Management Decision Support, 96
   in environmental decision-making, 167
   multicriteria analysis, 94
   simulation analysis, 94
   in soft systems methodologies, 95
   stochastic dynamic programming, 93
Decision analysis tools, 92–93
Decision authority, 136, 141–142
   and environmental laws, 142
   and general management statutes, 142
   need to unify among agencies, 145
Decision context, 130. *See also* Management context
   historical artifacts as obstacles, 160
Decision focus, single- *vs.* multifactor, 141
Decision makers, questions and issues for, 4
Decision-making authority, fragmentation of, 144
Decision-making procedure, 93, 94
Decision purpose, 136, 138–139
Decision scientists including, 146
Decision scope, 136, 139. *See also* Scope
   focus, 140–141
   goals, 141
   scale, 139
   Subic Bay redevelopment, 173–174
Decommissioning efforts, 23–24
Decomposers, 152, 154
Deep ecology movement, 17
   critiques of economic valuation from, 54–56
Degradation, by human activity, 119, 121
Delaney clause, 115
Deliberative ethics, 12, 19, 30
Deliberative processes, and stated preferences, 63
Delisting, of endangered species, 123
Delta smelt, 214–215
Derived values, 62
Developing countries, importance of addressing needs in, 47–48
Development, assessments of new, 25
Dewey, John, 12
Direct benefits, 136
Disclosure, as economic incentive, 47
Discounting, 84
Disruptors, 21
Distributional fairness, 16
Dose-response assessment, 133

Double counting, avoiding, 115
Drinking water, integration of ecology and economics, 118
DuPont Company, xiii
Dynamic ecological socioeconomic systems, 170
Dynamic stabilities, 103–104, 107
Dynamics, confronting in ecosystems, 100–107

E

Ecological analysis approaches, 86
Ecological assessment, integrating economics with, 45–46
Ecological attributes, 154
  in PCB-contaminated rivers valuation, 183–184
Ecological footprint, 70, 76, 91
  and complexity, 77
Ecological irreversibility, 159
Ecological properties
  ecologically relevant attributes, 153–154
  functional components of regional ecosystem, 151–152
  functional food web, 152
  identifying human values of, 154–156
  identifying relevant, 150–151
  properties common to all ecosystems, 151
  use of terminology, 147
Ecological relevance, 150
Ecological resources
  alternatives to monetary valuation of, xix
  incongruence with political jurisdictions, 18
  perceived value of, 3
  private sector valuation, 20–26
  rationale for assigning sociocultural values to, 16–17
  sociocultural valuation of, 9–10
Ecological risk assessment, 27, 98, 132
  in environmental decision-making, 2
  integrating socioeconomics and, 4–5
Ecological Risk Assessment Framework, 141
Ecological risk management (ERM) paradigm, 135
Ecological Society of America (ESA), xiv, 149
Ecological Sustainaiblity Index, 84
Ecological systems
  complexity in, 97–100
  confronting complexity, dynamics, variability and uncertainty in, 100–107
Ecological type, and levels of analysis, 102–103
Ecological valuation, 118, 147
  absence of simple methods for inclusion of, 165
  approaches to, 1–4
  comprehensive approach to identifying assessment endpoints, 146–147
  framework comparisons, 132–135
  influences of management context on integrating, 136–146
  informing process of, 160–161
  integrating socioeconomics and ecological risk assessment into, 4–5
  organizing and integrating process of, 129–131
  paradigm shift in, 5–7
  principles and characteristics of effective, 131–132
  private sector approaches, 20–26
  and public policy, 17–20
  revisions and time limits of, 171
  rudimentary understanding by risk assessors, 166
  value-based decision making, 131
Ecologically relevant attributes, 153–154
  in New Guinea rainforest ecosystem, 155
Ecologists
  conflicts with economists, 166
  questions and issues for, 4
Ecology, integration with economics, 107–115, 125
Economic effects data, 142
Economic efficiency, 46
Economic equity, 46
Economic impact analysis (EIA), 67, 75, 84, 137
  difficulties with heterogenicity, 77
  overlap with social impact assessment, 73
Economic incentives, 47
  and emissions trading, 19
Economic opportunity, in Madagascar case study, 120, 190–191
Economic valuation, 56, 60, 63, 86
  critiques from outside economics, 54–56
  for environmental decision-making, 48–50
  fairness issues, 50
  Kantian rejection of, 12
  theory-based critiques within economics, 50–52
  usefulness with bounding of ecological uncertainty, 125–126
  via revealed preference methods, 65
Economics
  in environmental decision-making, 2
  identifying endpoints for evaluation with, 157–159
  importance of including early in valuation process, 131
  integrating with ecological assessment, 45–46, 107–115, 125
  positive, 45
  roles in environmental policy analysis, 46–48
  as science of allocation of resources, 45
  as science of choosing, 45

Index

theory-based critiques of valuation from, 50–52
as untapped resource for environmental risk assessment, 47
Economist-ecologist joint valuation example, 115
additional considerations, 118
application of benefit-cost analysis, 116–117
available information, 116
problem statement, 116
uncertain future losses from extinction, 117–118
Economists
conflicts with ecologists, 166
and costs of remedial options, 157
ignorance of role of ecosystems, 115
inclusion in risk assessment, 171
questions and issues for, 4
value definitions by, 168
Ecosystem boundaries, transcendence of administrative boundaries, 144
Ecosystem Management Decision Support (EMDS), 96
Ecosystem quality, and CALFED Bay-Delta Program, 209
Ecosystem restoration, 212
Ecosystem services, valuation of, 118
Ecosystems
criteria for maintaining or restoring, 101
defining, 151
as types, 102
Ecoterrorism, 195
Ecotourism, 22
Effects, and type I/II errors, 105–106
Efficiency, and economic analyses, 16
Einstein, Albert, 107
Electrical power, valuation of, 69
Elicitation methods, 65
Embeddedness, 17
Emergy concept, 69
Emissions trading, 19, 47, 137
Emjoules, 69
Endangered species, 154
Chinook salmon, 208
and conflicts among stakeholders, 176
valuation methods, 66, 118
Endangered Species Act (ESA), 17, 19, 67, 142
and delta smelt, 214–215
Kantian ethics and, 12
Endocrine disruption mechanism, 80
Enduring values, 11, 54–56
Energy, as biophysical measure of value, 69
Energy and nutrient fixation, 154
Energy-based methods, 75
Energy flows, 95
Energy methods, 69

Enterprise architecture information technologies, 146
Environmental assets
difficulty of assigning economic values to, 50
stakeholder values based on proximity to, 169
Environmental damage schedules, 65
Environmental decision-making, xix
disconnect in, 114
economic valuation for, 48–50
frameworks for, 167
handling of incomplete knowledge in, 78–81
as hierarchical process, 17
inherent uncertainty in, 98
integrating ecology and economics to identify endpoints for, 157–159, 165
management context influences on integrating valuation into, 136–146
private sector examples, 21–26
in public sector, 19
stakeholders in, 1
value-based, 131
Environmental impact assessment (EIA), 1
cradle-to-grave approach to, 26
Environmental indices, 69–70
ecological footprint, 70
environmental sustainability index, 71–72
genuine progress indicator, 71
green accounting, 70
mass balance, 70
natural step, 70–71
Environmental laws, 142
Environmental performance index (EPI), 70
Environmental policy analysis, roles of economics in, 46–48
Environmental stochasticity, 103
Environmental sustainability index, 70, 71–72
Environmental valuation questionnaires, 43
Environmental values, 133
Environmental water
in CALFED case study, 122
difficulty of monitoring progress due to, 213–214
as new use category, 20
purchase of, 18
Equilibrium models, 69
Equity, economic, 46
Established indices, estimating percent loss using, 115
Evaluation criteria, 73
handling of incomplete knowledge in decision making, 78–81
heterogenicity capture, 74–77
incorporation of complexity, 77
ordinal vs. cardinal measures, 81–82
practicality, 83–84

Everglades National Park (ENP), 170
Ex post-facto damage assessment, 158
Expenditures, valuation methods focusing on, 67
Expert panels, 41–42
Exposure, seasonal variations in, 104
Exposure assessment, 133
Exposure doses, in PCB case study, 124
Extent, in scaling, 102
Externalities
  incorporation into decision-making, 61
  resolving through economics, 47
Extractive development. *See also* Madagascar case study
  Madagascar case study, 119–122
Exxon Valdez oil spill, 51

F

Fairness criterion, 50
Family-level impacts, 28
Federal trustees, as stakeholders, 181
Fetzer Bonterra wine, 22
Fetzer Wine Company, discontinuation of pesticide use by, 22
Field observation/monitoring, 104
Financial constraints, 29
Financial management experts, inclusion of, 146
Financial returns, present *vs.* future values, 116–117
Finite resources, 17
Fish, integration of ecology and economics, 113
Fixed damage schedules, 65
Focus groups, 32
  qualitative data collection through, 40–42
Food, analyzing, 154
Food and Drug Administration, carcinogen management regulations, 115
Food web
  for forest ecosystem, 153
  functional, 152
For-profit industries, challenges in assessing public concerns, 22
Forecasting, 104
Forest ecosystem
  ecologically relevant attributes in New Guinea, 155
  food web for, 153
  functional components, 153
Framework comparisons, 132
  review of existing environmental management frameworks, 134–135
  review of existing regulatory frameworks, 132–134
  synthesis, 135

Functional components
  identifying, 151
  of regional ecosystems, 151–152
  in tropical New Guinea rainforest ecosystem, 154
  vs. taxonomic groups, 152
Functional food web, 152
Funding problems
  in assessment of ecological endpoints, 147
  Gulf War oil spill, 195
  Subic Bay redevelopment, 177–178
Future benefits, unpredictability of, 213
Future period, in valuation studies, 116
Future preferences, 118
Future value, fallacy of basing on current preferences, 117
Fuzzy set theory, 84
  and uncertainty, 80

G

Game theory, role in economic analysis, 47
General management statutes, 142
Genetic toxins, impact on biological systems, xiii
Genuine progress indicator, 70, 76, 77, 84, 91
Global stewardship, 72
Goal setting, 146, 147–148
  CALFED Bay-Delta Program, 209–210
  Gulf War oil spill, 196
  in Madagascar case study, 188
  PCB-contaminated rivers valuation, 180–181
  Subic Bay redevelopment, 174
Golder Associates Ltd., xiii
Government Performance and Results Act (GPRA), 142
Government units, as stakeholders, 148
Grain, in scaling, 102
Grassroots for-profit ecotourism, 22
Great Lakes Ecological Assessment, 139
Greatest good for the greatest number, 11, 51
Greatest social good, 11
Green accounting, 70, 71, 91
Green GDP, 70, 75, 92
Green labeling, 47
Green products, consumer demand for, 136
Greening of business, 21
Groundwater, integration of ecology and economics, 108
Group-level values, 27
Groups, properties of, 13
Growth rate, estimating percent loss using, 114
Guardrails, 12
Gulf War oil spill, 40, 195
  analysis plan, 198
  chemical toxicity of oil residues, 198

Index

conceptual model, 197
data requirements, 197
hydrological functioning of tidal channels and, 199
lessons learned, 199
limited sources for recruitment of biota, 199
physical barriers affecting ecology, 199
physical toxicity of oil residues, 198–199
project characterization and communication, 198–199
remediation approach for shoreline habitat damaged by, 7
resource management goals and objectives, 196
scope, 195–196
stakeholder identification, 196–197

H

Habermas, Jürgen, 12
Habitat, 154
Habitat availability studies, 113
Habitat equivalency analysis (HEA), 67–69, 75, 84
  reducing speculation in, 114
Habitat loss, ecologist-economist uncertain valuation of, 115–118
Habitat types, identifying appropriate, 114
Harm, ranking of, 65
Hazard identification, 133
  limitations of current systems, 144
Heterogenicity, 75, 76
  ability to capture, 74–77
  over space, 74
  over time, 74
Heuristic rules of thumb, 80
Hierarchy theory, 100
Historical artifacts, 160
  in management context, 144
Holism, 140
Human context responsiveness, 82, 83
Human values, 130, 165
  as assessment endpoints, 156
  as basis of conducting ecological risk assessment, 143
  flux relative to gasoline prices, 169
  identifying in ecosystem components, 154–156
  outlooks on role in decision making, 167
Humanistic altruism, 10
Hypothetical choices, in surveys, 53

I

Ignorance, 79
  and element of surprise, 80
  and uncertainty, 78
  vs. uncertainty, 118

Ilmenite mine, 23
Impact assessment, Gulf War oil spill, 195–196
Incident response decisions, 139
Incomplete knowledge. *See also* Uncertainty
  handling of, 78–81
Indices, 76
  estimating percent loss using, 115
Indigenous peoples, 154
  addressing needs of, 47–48
  and limitations of market concepts of value, 51
  and Madagascar case study, 119
  and service flows in Subic Bay redevelopment, 175
Indirect benefits, 136
  challenges of measuring, 123
Individual goods, 51
Individual impacts, 28
Individual interviews, 32
  qualitative data collection through, 42–43
Individual preferences
  trade-offs of, 60
  vs. group well-being, 169
Individual values, 14
Inflationary Impact Statements, 2
Influence diagrams, 95
Information system design, 139
Informed decisions, 130, 160–161, 189
Infrastructure needs, 28
Injury, 68
  determination in PCB-contaminated rivers valuation, 183
  difficulty of quantifying, 114
  duration and severity over time, 159
Institutional capacity, 72
Integrated analysis approaches, 86, 171
  early in assessment process, 125
  in Madagascar case study, 193
Integrated solutions, in Madagascar case study, 121
Integration, 131
  identifying opportunities for better, 145–146
  influence of management context on, 136–146
  management context influences on, 143–146
Inter-Disciplinary Teams (IDTs), 142, 161
Interactive technology assessment, 72
International Association of Landscape Ecologists (IALE), xiv
International environmental management standards, 132
International scale decisions, 140
Interval analysis, 79
Interview scripts, 42, 43
Intrinsic values, 12, 54–56
Invertebrates, integration of ecology and economics, 111

Irreversible decisions, 117, 159
  avoiding, 118
Irreversible effects
  avoidance of, 81
  in hypothetical development case study, 116

J

Journal of Economic Perspectives, 53

K

Kant, Immanuel, 54–55
Kantian ethics, 12
  bright line in, 19
Kapustka, Lawrence A., xiii–xiv

L

Landscapes
  as defined spatial extent, 103
  as types, 102
Large-scale dams, change in social valuation of, 15
Lavaca Bay, 141, 150
Legal requirements
  and determination of values, 167
  in environmental decision-making, 2
Lemur, 188
  in Madagascar case study, 120
Leopold, Aldo, 102
Lessons learned
  appreciating/understanding complexity, 125–126
  Gulf War oil spill, 199
  integrating ecology with economics, 125
  Madagascar case study, 193–194
  in Madagascar case study, 121–122
  Subic Bay redevelopment, 177–178
Levee system integrity, CALFED Bay-Delta Program, 209
Level of analysis
  and conceptual models, 102–103
  seeking common, 100–101
Life cycle analysis, 72–73, 76
  sensitivity to context, 83
Life stages, sensitivity variation with, 104
Likelihoods, knowledge about, 79
Likert-type scale measures, 43
Littoral forest
  conceptual model, 189
  conservation of remnants, 188
  in Madagascar case study, 119–122
Local businesses, measuring gains to, 137

Local communities
  in PCB-contaminated rivers valuation, 181
  role in environmental decision-making, 137
Local human uses, conflicts with values of resources, 155–156
Local scale decisions, 140
Loss, 68
Lowest observed effect level (LOEL), estimating percent loss using, 115

M

Madagascar case study, 126, 187
  conceptual model, 189–190
  description of complexities, 119
  economic opportunity in, 190–191
  integrated solutions, 121
  lessons learned, 193–194
  natural environment in, 119–120
  problem statement, 187–188
  resource management goals, 188
  significant outcomes and lessons learned, 121–122, 193–194
  stakeholders in, 120, 188–189
  values identification, 191–192
Mammals, integration of ecology and economics, 112
Management context, 138
  decision authority, 141–142
  decision purpose, 138–139
  decision scope, 139–141
  elements of, 136
  influences on integrating valuation process, 136
  limitations and opportunities for improvement, 13–146
  and public/private decision-making, 136–138
Management reform statutes, of 1990s, 142
Mark-recapture experiments, 12
Market-like values, 48
Market share, negative effects on, 20
Market values, 52
Markets
  favoring by economists, 48
  as source of data on trade-offs, 61
Mass balance, 70–71, 71, 76
Material flows, 95
Materiality, estimating percent loss using, 114
Maximin, 80
Maximum likelihood methods, 106
Media, stakeholder role in PCB-contaminated rivers valuation, 181
Methodological individualism, 14
Mineral mining, ecologically sensitive approaches for, 7

Index

Minimax regret decision criteria, 80
Minnesota IMPLAN Group, 67
Mitigation planning, Subic Bay redevelopment, 176
Model estimation, 106
Monetary valuation, 61, 63
  alternatives to, xix
  for benefits from Clean Air Act, 203
  nonrequirement of, 126
  problems in GPI, 71
  requirements for, 2
Monetization, 82
  and Clean Air Act benefits valuation, 204–205
  PCB-contaminated rivers valuation, 180
  problematic absence of, 168
Monitoring programs, 73
Montreal Process criteria, 142
Moral theories, 55
  relevance to valuation, 54
Multicriteria analysis, 65, 75, 77, 82, 94
Multifactor decision focuses, 140, 141, 153
Multiple-use land and resource management plans, 142
Multiplier effects, of business activity, 137
Multiscale environmental indicators, 143
Multiscale linkage of decisions, 139, 143
Munns, Wayne R., Jr., xiv

N

National Environmental Policy Act of 1969 (NEPA), 141
National Forest Management Act, 142
National Hierarchical Framework of Ecological Units, 145
National Research Council (NRC), 3, 31, 41
National scale decisions, 140
Natural capital, 68
Natural context responsiveness, 82, 83
Natural entities, nonfungibility with money, 55
Natural environment, in Madagascar case study, 119–120
Natural income, 68
Natural resource damage assessment and restoration (NRDAR), 114, 124, 171
  injury determination in PCB-contaminated rivers valuation, 183
  PCB-contaminated rivers valuation, 180–181
  public engagement with, 184
  relationship to CERCLA, 185
Natural resource damage assessments, 68
Natural resource management, 18
  successes and failures, 19–20
  valuation based on, 10
Natural step, 70–71, 71, 76
Natural System conceptual model, 5

Natural Systems
  ability to incorporate complexity in, 77
  heterogenicity limitations, 77
Nature Conservancy, The, 21–22
Nesting of communities, 15
Net benefits, present value of, 117
Net Environmental Benefit Assessment (NEBA), 68
NetWeaver knowledge base system, 96
Networks of social relationships, 12, 13
New chemicals, registration of, 17
New Guinea
  ecologically relevant attributes of functional components, 155
  functional components of tropical rainforest ecosystem, 154
No observed effect level (NOEL), estimating percent loss using, 115
Nominal group technique, 41
Nonchemical stressors, 134
  incorporating into valuations, 133, 135
Noneconomic valuation, 60
Nongovernmental organizations (NGOs), 41, 136
  role in Madagascar case study, 120
  as stakeholders, 148
Nonmarket valuation, 33, 44, 52
Normative economics, 45–46
Numeral unit spread assessment pedigree (NUSAP), 79

O

Observational boundaries, 102
  grain and extent, 102
Observer dependence, of landscape boundaries, 103
Occam's razor, 100
Occupational health hazards, 81
Office of Solid Waste and Emergency Response (OSWER), 133, 134
Open appraisal approaches, 61–62
Open decisional frameworks, 171
Openness to change, 10
Opportunity cost method, 66–67, 75
Option value, 117
Ordinal measures, 81–82
Organisms, as type, 102
Outcomes
  failure to coordinate with economists, 114
  knowledge about, 79
  in Madagascar case study, 121–122
  ranking with ordinal measures, 81
  selection of preferred, 44
Outside experts, in Madagascar case study, 193
Ozone depletion, 81

P

Pacific Northwest Forest Plan, 139
Paired comparison methods, 65
Paradigm shift, in ecological valuation, 5–7
Pareto criterion, limitations in environmental policy decisions, 48–49
Participatory processes, 30, 132
   Subic Bay redevelopment, 177
Participatory technology assessment, 72
PCB-contaminated rivers, 124, 179. *See also* Polychlorinated biphenyls (PCBs)
   analysis plan, 182–184
   assessment endpoint, 182
   characterization and communication, 184
   conceptual model, 181–182
   data requirements, 182
   remediation of, 184
   resource management goals and objectives, 180–181
   scope of project, 179–180
   stakeholder identification, 181
   values identification, 182
Pellston Workshop on Valuation of Ecological Resources, xxi, 5, 149
Percent loss of services, 114
   from increased risk of extinction, 117
Persistent effects, avoidance of, 81
Pervasive uncertainty, 118
Pesticide use, discontinuation by Fetzer Wine Company, 22
Philippines. *See* Subic Bay redevelopment
Physical resources, 176
Physical scale, 101
Physiology, values and, 13
Planning decisions, 139
Planning entities, as stakeholders, 148
Plug and play tools, for valuation methods, 84
Polanski, Roman, 207
Policy decisions, uncertainty in, 98
Political bias
   in environmental decisions, 169
   in Subic Bay redevelopment, 178
Political jurisdictions, incongruence with ecological systems, 15, 18
Politics, in environmental decision-making, 2
Pollination, 154
Pollution reduction, economic incentives for, 47
Polychlorinated biphenyls (PCBs)
   acceptable limits for, 2
   case study, 124
   remediation and restoration of river contaminated by, 7

Population
   Chinook salmon, 209, 212
   as type, 102
   and variability, 105
Population impacts, 28
Population modeling, integration of ecology and economics, 113
Population resilience, 123
Positive economics, 45
Postinjury state, 159
Potential harm, 116
Potential Pareto improvement criterion, 49
Potentially responsible party (PRP), 181
   in Gulf War oil spill, 195
Poverty amelioration, in Madagascar case study, 191
Practicality, 75, 76
   as evaluation criterion, 83–84
Precautionary principles, 12
Predation, 154
Preferences
   change over time, 169
   current *vs.* future, 117–118
   inconsistencies among groups, 169
   stability and compensation test validity, 50
   as starting point in economics, 54
   vs. aesthetic judgments, 55
   vs. enduring values, 11
   vs. values, 118
Present value
   of loss/net benefits, 117
   of waiting to decide, 117
Presidential/Congressional Commission on Risk Management, 135
Primary data collection, 84
Private sector
   business impacts valuation problem, 21
   examples of decision-making, 21–26
   land management in site decommissioning, 23–25
   prospects for valuation in, 26
   role in decision-making, 136–137
   valuation approaches in, 20–21
Probabilistic risk assessment
   limits of, 79
   in PCB case study, 124
   and uncertainty, 78
Probability density functions, and uncertainty, 78
Problem formulation, 133
Problem identification, 146, 147–148, 148
Problem statement, Madagascar case study, 187–188
Producers, 152, 154
Product acceptability, negative effects on, 20
Profitability, negative effects on, 20

Index

Program design, 138
Program evaluation specialists, inclusion of, 146
Programmatic Record of Decision, 210
Project-level determinations, 139
Project successes, Subic Bay redevelopment, 178
Propagule dispersal, 154
Property locations, and attributes of, 66
Property rights
  and contractarianism, 52
  and nonmarket valuation, 52
Protection levels, determining, 3
Proximity, to environmental resources, 169
Psychology, values and, 13
Public decision-making, 136, 137–138, 138
  in cleanup of Superfund sites, 167
  PCB-contaminated rivers valuation, 184
Public goods, 47
Public health hazards, 81
Public policy
  ecological values and, 17–19
  and natural resource management successes and failures, 19–20
  and public sector decision-making mechanisms, 19
Public sector, decision-making mechanisms in, 19
Public trust, as assessment endpoint, 167
Public values, in environmental decision-making, 2

Q

QIT Madagascar Minerals (QMM), 120, 187, 188, 190
Qualitative data, 31–32, 172
  collection through individual interviews, 42–43
  focus groups and, 40–42
  methods for eliciting, 32, 40–43
  from scientists, 168
Quantitative data, 31–32, 172
  attitude surveys for collecting, 43
  methods for eliciting, 33, 43–44
  valuation surveys for collecting, 43–44
Quasi-option value, 117
Questionnaires, 31, 43

R

Random utility theory, 64
Ranking of harm, 65
Rates of exchange, 101
Recreation enhancements, 183–184
Reflexive process, 27
Regional Economic Models, Inc. (REMI), 67

Regional ecosystems, functional components of, 151–152
Regional planning process, 121, 122
Regional scale decisions, 140
Regulatory agencies, as stakeholders, 148
Regulatory frameworks, 132–134
Regulatory hindrances, 115
Regulatory impact analyses (RIAs), 49–50
Regulatory requirements, 19
  in environmental decision-making, 2
Replacement cost method, 66–67, 75
  and complexity, 77
Representative species, identifying appropriate, 114
Reproductive rate, estimating percent loss using, 114
Researchers' notes, qualitative data from, 31
Resident-newcomer conflicts, 28
Resilience, 123
  difficulty of modeling, 98
Resource allocation, 139
  in Madagascar case study, 193–194
Resource Equivalency Analysis (REA), 67
  reducing speculation in, 114
Resource management
  CALFED Bay-Delta Program, 209–210
  Gulf War oil spill, 196
  in Madagascar case study, 188
  PCB-contaminated rivers valuation, 180–181
  Subic Bay redevelopment, 173–174
Resources
  allocation of, 45
  integration of ecology and economics for, 108–113
Response rates, maximizing, 33
Responsible economic development, in Madagascar case study, 191
Responsible party. *See* Potentially responsible party (PRP)
Restoration baseline, in HEA method, 68
Restoration benefits, in HEA method, 68
Restoration scaling, in PCB-contaminated rivers valuation, 183
Revealed preference methods, 52, 65–66, 75
  and complexity, 77
  heterogenicity capture with, 77
Rights-based ethics, 55
Rio Tinto mine development, 23, 25, 120, 187
Risk, *vs.* uncertainty, 118
Risk assessment (RA), 1, 79, 98, 159
  integration of ecology and economics, 113
  role of economics in, 47
  separation of uncertainty and variability in, 105
Risk characteristics, 133

Risk condition, and uncertainty, 78
Risk management decisions, 17
   inputs to, 2
Risk managers, xix
Risk tolerance, 3
Robustness, against natural and evaluative error, 86
Runoff reduction, 183

S

Sacramento River Delta, 7, 67, 208
Safe minimum standards, 12
Sampling, 74
Sampling error, avoiding, 33
Scale, 102
   in decision scope, 139
   and level of analysis, 102–103
   observer independence of, 101
Scaling, 68
Scaling relationships, 100
Scenario analysis, 62, 79, 80
Scenario building, 104, 126
Science, definition, 45
Science Advisory Board (SAB), 133, 166
Scope. *See also* Decision scope
   Gulf War oil spill assessment, 195–196
   PCB-contaminated rivers valuation, 179–180
Sediment, integration of ecology and economics, 110
Self-interest, values based on, 10
Self-organization, in ecological systems, 98
Sensitivity
   issues in CALFED case study, 123
   and life stage, 104
   and type II errors, 106
Sensitivity analysis, 62, 79, 80, 84
Service flows
   for indigenous communities, Subic Bay redevelopment, 175
   in PCB-contaminated rivers valuation, 184
   and valuation of Clean Air Act, 202
Services
   percent loss of, 114
   usefulness of focus on, 130–131
Shareholder value, negative effects on, 20
Shoreline development. *See also* Gulf War oil spill
   TNC involvement in, 22
Shrimp mariculture, 69
Sierra Nevada Framework Assessment, 139
Simulation analysis, 94
Single-factor decision focuses, 141
Site decommissioning, 23–25
Site remediation, 18
Site-specific dose-response studies, 113

Slash and burn agriculture, 120
Small-group discussions, 63
Small-scale environmental changes, difficulty of valuing, 158
Social capacity, 72
Social factors, 27
   in environmental decision-making, 2
Social goods, defining nature of, 51
Social impact assessment, 73, 76
Social impacts, 27
   assessment of, 28
Social indicators
   in data collection, 28
   identification of relevant, 29
Social issues, Subic Bay redevelopment, 176
Social psychology
   of group dynamics, 41
   and valuation, 10
Social System conceptual model, 5
Societal values, 155
Society of Environmental Toxicology and Chemistry (SETAC), xiii, xiv, xix, 149
   Pellston Workshop on Valuation of Ecological Resources, xxi, 4, 5, 41
   Valuation of Ecological Resources workshop, 172
Sociocultural indices, 69–70
Sociocultural values
   and economic values, 16
   rationale for assigning to ecological resources, 16–17
Socioeconomics
   conditions requiring determination of values, 167
   integrating with ecological risk assessment, 4–5, 6, 165
Soft systems methodologies, 95
Soil, integration of ecology and economics, 109
South Florida Ecosystem Restoration Plan, 139
Space, heterogeneity over, 74
Spatial dimensions, 99
Spatial scale, 100, 158
   of environmental decisions in valuation context, 140
Species extinction, 116
   avoiding in CALFED case study, 123
   economist-ecologist uncertain valuation of, 115–118
   high valuation of, 67
   irreversibility of, 159
   potentials for Chinook salmon, 212
   uncertain future losses from, 117–118
Stable system states, 95
Staffing considerations, in Madagascar case study, 193

Stahl, Ralph G., Jr., xiii
Stakeholder empowerment, 178
  in Madagascar case study, 191, 193, 194
Stakeholder identification, 131, 148–149
  CALFED Bay-Delta Program, 210–211
  Gulf War oil spill, 196–197
  Madagascar case study, 188–189, 189
  PCB-contaminated rivers valuation, 181
  Subic Bay redevelopment, 174
Stakeholder maps, 40
Stakeholders, 30
  alternative livelihood strategies, 178
  challenges in deciphering views, 3
  difficulty of satisfying all, 169
  in environmental decision making, 1
  inclusion in decision-making process, 134
  in Madagascar case study, 120
Standards, setting, 17
Stated preference methods, 53, 63, 75
  attribute-based methods, 64
  choice experiments, 64
  contingent valuation, 63–64
  environmental damage schedules, 65
  sensitivity to context, 83
  variants on elicitation methods, 65
States of knowledge, 78
Statistical design, of elicitation approaches, 64
Stewardship, 16, 17, 25, 72
  as assessment endpoint, 167
  in Madagascar case study, 120
Stochastic dynamic programming, 93
Stock prices, 66
Stocks, measuring capital assets as, 62
Strategic bias, 83
Strategic direction, at top management level, 138
Strategic manipulation, and context dependence, 82
Stressor levels, 158
Stressor response, 133
Structure and function, of ecological properties, 151
Subic Bay redevelopment, 7, 48, 51, 52, 166
  analysis plan, 177
  characterization and communication, 177
  conceptual model, 174–175
  data requirements, 176–177
  goals and objectives, 174
  lessons learned, 177–178
  resource management, 173
  scope of, 172–174
  stakeholder identification, 174
  values identification, 175–176
Subsistence agriculture, in Madagascar case study, 120
Substitutability, 55

Sulfur dioxide allowances, 47
  tradable, 48
Superfund, 124, 150
  and Gulf War oil spill, 197
  public role in cleanup decisions, 167
  risk assessment guidance for, 147
Superfund Amendments of 1986, 133
Supply and demand, 45
Surface water, integration of ecology and economics, 108
Surprise, 118
  and uncertainty/ignorance, 80
Surveys, 33, 42, 43, 63
  hypothetical choices in, 53
Sustainability guidelines, 70, 71

T

Technical analysis, as basis for policy making, 62
Technological feasibility, in environmental decision-making, 2
Technology assessment, 72, 76
  sensitivity to context, 83
Temporal dimensions, 99
Temporal scale, 100, 158
  of environmental decisions in valuation context, 140
Terminology problems, among fragmented agencies, 145
Theory of utility from attributes, 63
Threatened species, 154
Timber Assessment Market Model, 203
Timber harvesting, conflicts among stakeholders, 176
Time, heterogenicity over, 74
Tools, for valuation methods, 84
Tourism, in Everglades National Park, 170
Tractability, constraints and, 103
Tradable emission permits, 61
Tradable Environmental Allowances (TEAs), 19
Trade-off analyses, 106–107, 126, 213
  among risks, 159
  CALFED Bay-Delta Program, 211
  in CALFED case study, 122
  in Madagascar case study, 122
Trade-offs, 44
  among services, 60
  in CALFED Bay-Delta Program, 210
  concerning environmental goods and services, 11
  in decision-making, 93
  in environmental decision-making, xix, 1
  FDA example, 2
  inferring from data on choices, 65–66
  measures of weights between attributes, 64

Traditional values, 10
Transaction analysis, 15
Transboundary circumstances, 144–145
Transcripts, qualitative data from, 31
Transparency, in ecological management, 132
Transport processes, 151
Trophic categories, 152, 154
Type, 102
Type I errors, 105–106
Type II errors, 106

U

Uncertain future losses, from species extinction, 117–118
Uncertain valuation, ecologist-economist example, 115–118
Uncertainty, 75, 79
  analyzing separately from variability, 105
  Bayesian analysis for, 94
  condition of, 78
  confronting in ecosystems, 100–107
  degree associated with alternative outcomes, 159
  handling in decision making, 78–81
  inherence in policy decisions, 98
  in managing complex systems, 103–106
  pervasive, 118
  robust methodological responses to, 79
  trade-off analysis and, 106
  type I and type II errors from, 105–106
  vs. ignorance, 118
Unintended consequences, 5
United Nations Compensation Commission (UNCC), 196
US Environmental Protection Agency (USEPA), xiii, 17, 133, 201
  Atlantic Ecology Division, xiv
  Guidelines for Ecological Risk Assessment, 134
  Office of Research and Development, xv
  Risk Assessment Guidance for Superfund, 147
  role in PCB-contaminated rivers valuation, 181
  Science Advisory Board, 166
US Geological Survey, Hydrologic Unit Codes, 145
US Office of Management and Budget (USOMB), 142
Use categories, creating new, 20
Use sectors, conflicts between public and private, 20
USEPA Framework, 134
USEPA Science Advisory Board, 135
Utilitarianism, 51, 56
  and economic valuation, 46
  Kant's opposition to, 54–55

V

Valuation methods, 59–61, 62
  and absence of monetization, 168
  criteria for, 60
  economic impact analysis, 67
  energy methods, 69
  environmental and sociocultural indices, 69–72
  evaluation criteria, 73–84
  habitat equivalency analysis, 67–69
  life cycle analysis, 72–73
  open and closed appraisal approaches, 61–62
  opportunity cost, 66–67
  plug and play tools for, 87
  replacement cost, 66–67
  responsiveness to context, 82–83
  revealed preference, 65–66
  social impact assessment, 73
  stated preference, 63–65
  technology assessment, 72
Valuation process. See Ecological valuation
Valuation studies, 48
Valuation surveys, 43–44
Value
  changes over time, 126
  defining nature of, 6, 10–12, 168
  economic definition, 46
  vs. preferences, 118
Value-based decision making, 131
Values of interest, 12–14
  CALFED Bay-Delta Program, 212
  change over time, 15, 169
  Gulf War oil spill, 197
  importance of context, 14–15
  individual and community-based, 14
  Madagascar case study, 191–192
  PCB-contaminated rivers valuation, 182
  Subic Bay redevelopment, 175–176
Variability, 104, 107
  analyzing separately from uncertainty, 105
  confronting in ecosystems, 100–107
  consequences of ecosystem, 105
  in managing complex systems, 103–106
  reduction of sensitivity through, 106
  of seasonal exposure, 104
  trade-off analysis and, 106
Vegetation, integration of ecology and economics, 109
Vertebrates, 111
Video recordings
  of focus group sessions, 41
  qualitative data from, 31
Virginia Coast Reserve, 21–22

W

Water curtailments, ecological benefits of, 122
Water quality ladder, 158, 209
Water supply, and CALFED Bay-Delta Program, 209
Watermen, in Chesapeake Bay blue crab industry, 18–19
Watershed conditions, 145
  in CALFED Bay-Delta Program, 209
Welfare economics, 48
Western philosophy, utilitarianism in, 46
Western states, water allocation issues, 20
Wetland services
  restoration of, 183
  valuation methods, 66
Wetlands, integration of ecology and economics, 110
Wicked problems, 98
Wildfires, attitudes based on human values, 130
Willingness to accept (WTA), 63
  limitations of, 54
Willingness to pay (WTP), 16, 49, 82, 138
  limitations of, 54
Winners and losers, 73
  in compensation tests, 49
World Bank, 119, 188
World Wide Fund for Nature, 119

Other Titles from the Society of Environmental Toxicology and Chemistry (SETAC):

Atrazine in North American Surface Waters: A Probabilistic Aquatic Ecological Risk Assessment
Giddings, editor
2005

Effects of Pesticides in the Field
Liess, Brown, Dohmen, Duquesne, Hart, Heimbach, Kreuger,
Lagadic, Maund, Reinert, Streloke, Tarazona
2005

Human Pharmaceuticals: Assessing the Impacts on Aquatic Ecosystems
Williams, editor
2005

Toxicity of Dietborne Metals to Aquatic Organisms
Meyer, Adams, Brix, Luoma, Stubblefield, Wood, editors
2005

Toxicity Reduction and Toxicity Identification Evaluations for Effluents, Ambient Waters, and Other Aqueous Media
Norberg-King, Ausley, Burton, Goodfellow, Miller, Waller, editors
2005

Use of Sediment Quality Guidelines and Related Tools for the Assessment of Contaminated Sediments
Wenning, Batley, Ingersoll, Moore, editors
2005

Life-Cycle Assessment of Metals
Dubreuil, editor
2005

Working Environment in Life-Cycle Assessment
Poulsen and Jensen, editors
2005

Life-Cycle Management
Hunkeler, Saur, Rebitzer, Finkbeiner, Schmidt,
Jensen, Stranddorf, Christiansen
2004

Scenarios in Life-Cycle Assessment
Rebitzer and Ekvall, editors

Ecological Assessment of Aquatic Resources: Linking Science to Decision-Making
Barbour, Norton, Preston, Thornton, editors
2004

Life-Cycle Assessment and SETAC: 1991–1999
15 LCA publications on CD-ROM
2003

Amphibian Decline: An Integrated Analysis of Multiple Stressor Effects
Greg Linder, Sherry K. Krest, Donald W. Sparling
2003

Metals in Aquatic Systems:
A Review of Exposure, Bioaccumulation, and Toxicity Models
Paquin, Farley, Santore, Kavvadas, Mooney, Winfield, Wu, Di Toro
2003

Silver: Environmental Transport, Fate, Effects, and Models:
Papers from Environmental Toxicology and Chemistry, 1983 to 2002
Gorusch, Kramer, La Point
2003

Code of Life-Cycle Inventory Practice
de Beaufort-Langeveld, Bretz, van Hoof, Hischier,
Jean, Tanner, Huijbregts, editors
2003

Contaminated Soils: From Soil–Chemical Interactions
to Ecosystem Management
Lanno, editor
2003

Environmental Impacts of Pulp and Paper Waste Streams
Stuthridge, van den Heuvel, Marvin, Slade, Gifford, editors
2003

Life-Cycle Assessment in Building and Construction
Kotaji, Edwards, Shuurmans, editors
2003

Porewater Toxicity Testing: Biological, Chemical,
and Ecological Considerations
Carr and Nipper, editors
2003

Reevaluation of the State of the Science for Water-Quality
Criteria Development
Reiley, Stubblefield, Adams, Di Toro, Erickson,
Hodson, Keating Jr, editors
2003

Community-Level Aquatic System Studies—Interpretation Criteria (CLASSIC)
Giddings, Brock, Heger, Heimbach, Maund, Norman,
Ratte, Schäfers, Streloke, editors
2002

Interconnections between Human Health and Ecological Variability
Di Giulio and Benson, editors
2002

SETAC
A Professional Society for Environmental Scientists and Engineers and Related Disciplines Concerned with Environmental Quality

The Society of Environmental Toxicology and Chemistry (SETAC), with offices currently in North America and Europe, is a nonprofit, professional society established to provide a forum for individuals and institutions engaged in the study of environmental problems, management and regulation of natural resources, education, research and development, and manufacturing and distribution.

Specific goals of the society are:

- Promote research, education, and training in the environmental sciences.
- Promote the systematic application of all relevant scientific disciplines to the evaluation of chemical hazards.
- Participate in the scientific interpretation of issues concerned with hazard assessment and risk analysis.
- Support the development of ecologically acceptable practices and principles.
- Provide a forum (meetings and publications) for communication among professionals in government, business, academia, and other segments of society involved in the use, protection, and management of our environment.

These goals are pursued through the conduct of numerous activities, which include:

- Hold annual meetings with study and workshop sessions, platform and poster papers, and achievement and merit awards.
- Sponsor a monthly scientific journal, a newsletter, and special technical publications.
- Provide funds for education and training through the SETAC Scholarship/Fellowship Program.
- Organize and sponsor chapters to provide a forum for the presentation of scientific data and for the interchange and study of information about local concerns.
- Provide advice and counsel to technical and nontechnical persons through a number of standing and ad hoc committees.

SETAC membership currently is composed of more than 5000 individuals from government, academia, business, and public-interest groups with technical backgrounds in chemistry, toxicology, biology, ecology, atmospheric sciences, health sciences, earth sciences, and engineering.

If you have training in these or related disciplines and are engaged in the study, use, or management of environmental resources, SETAC can fulfill your professional affiliation needs.

All members receive a newsletter highlighting environmental topics and SETAC activities, and reduced fees for the Annual Meeting and SETAC special publications.

All members except Students and Senior Active Members receive monthly issues of *Environmental Toxicology and Chemistry (ET&C)* and *Integrated Environmental Assessment and Management (IEAM)*, peer-reviewed journals of the Society. Student and Senior Active Members may subscribe to the journal. Members may hold office and, with the Emeritus Members, constitute the voting membership.

If you desire further information, contact the appropriate SETAC Office.

1010 North 12th Avenue	Avenue de la Toison d'Or 67
Pensacola, Florida 32501-3367 USA	B-1060 Brussels, Belgium
T 850 469 1500 F 850 469 9778	T 32 2 772 72 81 F 32 2 770 53 86
E setac@setac.org	E setac@setaceu.org

www.setac.org
Environmental Quality Through Science®